Drug Targets in Kinetoplastid Parasites

ADVANCES IN EXPERIMENTAL MEDICINE AND BIOLOGY

Recent Volumes in this Series

Volume 616
TRANSGENIC MICROALGAE AS GREEN CELL FACTORIES
Edited by Rosa León, Aurora Gaván, and Emilio Fernández

Volume 617
HORMONAL CARCINOGENESIS V
Edited by Jonathan J. Li

Volume 618
HYPOXIA AND THE CIRCULATION
Edited by Robert H. Roach, Peter Hackett, and Peter D. Wagner

Volume 619
CYANOBACTERIAL HARMFUL ALGAL BLOOMS: STATE OF THE SCIENCE
AND RESEARCH NEEDS
Edited by H. Kenneth Hudnell

Volume 620
BIO-APPLICATIONS OF NANOPARTICLES
Edited by Warren C.W. Chan

Volume 621
AXON GROWTH AND GUIDANCE
Edited by Dominique Bagnard

Volume 622
OVARIAN CANCER
Edited by George Coukos, Andrew Berchuck, and Robert Ozols

Volume 623
ALTERNATIVE SPLICING IN THE POSTGENOMIC ERA
Edited by Benjamin J. Blencowe and Brenton R. Graveley

Volume 624
SUNLIGHT, VITAMIN D AND SKIN CANCER
Edited by Jörg Reichrath

Volume 625
DRUG TARGETS IN KINETOPLASTID PARASITES
Edited by Hemanta K. Majumder

Drug Targets in Kinetoplastid Parasites

Edited by
Hemanta K. Majumder

*Molecular Parasitology Laboratory, Indian Institute of Chemical Biology,
Kolkata, India*

Springer Science+Business Media, LLC
Landes Bioscience

Springer Science+Business Media, LLC
Landes Bioscience

Printed in the U.S.A.

Springer Science+Business Media, LLC, 233 Spring Street, New York, New York 10013, U.S.A.
http://www.springer.com

Please address all inquiries to the Publishers:
Landes Bioscience 1002 West Avenue, 2nd Floor, Austin, Texas, U.S.A. 78701
Phone: 512/ 637 6050; FAX: 512/ 637 6079
http://www.landesbioscience.com

Drug Targets in Kinetoplastid Parasites edited by Hemanta K. Majumder, Landes Bioscience /
Springer Science+Business Media, LLC dual imprint / Springer series: Advances in Experimental
Medicine and Biology

ISBN: 978-0-387-77569-2

While the authors, editors and publisher believe that drug selection and dosage and the specifications
and usage of equipment and devices, as set forth in this book, are in accord with current recommend-
ations and practice at the time of publication, they make no warranty, expressed or implied, with
respect to material described in this book. In view of the ongoing research, equipment development,
changes in governmental regulations and the rapid accumulation of information relating to the biomedical
sciences, the reader is urged to carefully review and evaluate the information provided herein.

Library of Congress Cataloging-in-Publication Data

Drug targets in kinetoplastid parasites / edited by Hemanta K. Majumder.
    p. ; cm. -- (Advances in experimental medicine and biology ; v. 625)
 Includes bibliographical references and index.
 ISBN 978-0-387-77569-2
 1. Antiprotozoal agents. 2. Drug targeting. 3.
Leishmaniasis--Chemotherapy. 4. Kinetoplastida. I. Majumder, Hemanta K.
II. Series.
 [DNLM: 1. Leishmaniasis--drug therapy. 2. Membrane Glycoproteins--drug
effects. 3. Protozoan Proteins--drug effects. W1 AD559 v.625 2008 / WR 350
D794 2008]
 RM413.D78 2008
 615'.7--dc22

                                                          2007049045

DEDICATION

This book is dedicated to the memory of my teacher late Prof. Radha Kanta Mandal, Ex-Chairman, Department of Biochemistry, Bose Institute, Kolkata, India and to the memory of late Prof. Amar Nath Bhaduri, a pioneer in *Leishmania* research in India and Ex-Director of Indian Institute of Chemical Biology, Kolkata, India.

PREFACE

If viewed globally, the parasitic diseases pose an increasing threat to human health and welfare. The diseases caused by kinetoplastid protozoan parasites like *Leishmania* and *Trypanosoma* continue to a cause suffering for many millions of people in both tropical and subtropical regions of the world. *Leishmania* species are found throughout Latin America, Africa and Asia. *Trypanosoma cruzi* that cause Chagas' disease is endemic in Latin America, while members of *Trypanosoma brucei* group are found in sub-Saharan Africa. Although the past two decades has witnessed commendable research efforts and technical advances in our understanding of the biochemistry and molecular and cell biology of these pathogens, the dreaded protozoal diseases caused by these organisms continue to threaten mankind. Therapeutic tools for the treatment of most parasitic diseases are extremely limited. The development of parasites resistant to many of the available drugs is also responsible for the depressing picture of disease persistence and death. Development of commercially available vaccines is still far from reality, though research and trial programs continue. The advent of genomics has increased this opportunity, and perhaps data derived from genome projects will hasten the development of effective, inexpensive, easy to administer vaccines against these parasites.

Kinetoplastid protozoan parasites have special features. They are characterized by the presence of a massive intercatenated network structure of DNA called the kinetoplast DNA, or kDNA, harbored within the single mitochondrion or the kinetoplast of these organisms. This kDNA is unique as none of the host organisms of these parasites contains DNA that resembles this DNA. Therefore, it can be an excellent target for developing therapeutic agents. This inhibition should result in selective killing of the parasite as the kinetoplast gene products are required at all stages of the life cycle of *Leishmania* and *Trypanosoma cruzi*.

Current biomedical research always has its focus in the search for newer intervention strategies to control the public health impact of parasitic diseases. The dramatic advances in molecular biology and genomics and proteomics have provided opportunities for discovering and evaluating molecular targets for drug design which now form a rational basis for the development of improved antiparasitic therapy.

When I was invited by Ron Landes of Landes Bioscience to edit a book on drug targets in kinetoplastid parasites, I accepted the offer as I felt that it is the right time to address this important subject. The book contains 12 chapters contributed by eminent scientists working in this field. The articles deal mainly with two aspects: visual identification of targets and identification of therapeutic agents.

Several targets like kDNA replication machinery, purine salvage pathway, purine and pyrimidine biosynthetic pathway, histone deacetylase, DNA topoisomerases, membrane transporter proteins, glycoproteins and glycolipids are discussed in this book.

Since current treatments for kinetoplastid parasitic diseases are far from ideal, there is an urgent and genuine need to develop newer compounds as antiparasitic drug candidates. Therefore development of some lead compounds against these parasites as well as drug resistance are also included in this book. Moreover the vast amount of information generated after publication of the "Trytrip" genome sequence now makes possible several new approaches for target identification and discovery of therapeutic agents.

This book is an outcome of the contributions of many scientists working in this important area. I am thankful to all of them. Finally, this book was made possible only because of continuous help from my Ph.D. student Mr. Agneyo Ganguly and the secretarial assistance from Mrs. Moumita Majumder.

Hemanta K. Majumder, Ph.D.

ABOUT THE EDITOR...

HEMANTA K. MAJUMDER is a Director Grade Scientist and Head of the Infectious Diseases and Immunology Division of Indian Institute of Chemical Biology, Kolkata, India. He is also the Working Chairman of the State Council of Science and Technology, Government of West Bengal, India. His main research interests include the biochemistry of DNA topoisomerases of *Leishmania* in relation to development of therapeutics and understanding the mechanism of programmed cell death in this unicellular organism.

Dr. Majumder is a reviewer of many peer-reviewed journals. He received his Ph.D. in Biochemistry from University of Calcutta in 1975 and did his postdoctoral studies at the Albert Einstein College of Medicine, New York, USA and the University of Zurich, Switzerland. He was a Visiting Associate and Research Molecular Biologist at the University of California at Berkeley, USA.

Dr. Majumder is a Fellow of the Indian National Science Academy (FNA), Indian Academy of Sciences (FASc) and National Academy of Sciences (FNASc), India.

PARTICIPANTS

Cassandra S. Arendt
Postdoctoral Fellow
Department of Biochemistry
  and Molecular Biology
Oregon Health & Science University
Portland, Oregon
USA

Sumi Mukhopadhyay nee
  Bandyopadhyay
Immunobiology Division
Indian Institute of Chemical Biology
Kolkata
India

Rachel Bezalel
Department of Parasitology
The Kuvin Center for the Study
  of Infectious and Tropical Diseases
The Hebrew University-Hadassah
  Medical School
Jerusalem
Israel

Jan M. Boitz
Department of Biochemistry
  and Molecular Biology
Oregon Health & Science University
Portland, Oregon
USA

Frederick S. Buckner
Department of Medicine
University of Washington
Seattle, Washington
USA

Nicola S. Carter
Department of Biochemistry
  and Molecular Biology
Oregon Health & Science University
Portland, Oregon
USA

Benu Brata Das
Department of Molecular Parasitology
Indian Institute of Chemical Biology
Kolkata
India

Alok Kumar Datta
Division of Infectious Diseases
Indian Institute of Chemical Biology
Jadavpur, Kolkata
India

Rupak Datta
Division of Infectious Diseases
Indian Institute of Chemical Biology
Jadavpur, Kolkata
India

Chinmoy S. Dey
Department of Biotechnology
National Institute of Pharmaceutical
  Education and Research
Punjab
India

Agneyo Ganguly
Department of Molecular Parasitology
Indian Institute of Chemical Biology
Kolkata
India

James W. Goding
Walter and Eliza Hall Institute
    of Medical Research
Department of Physiology
Monash University
Victoria
Australia

Emanuela Handman
Walter and Eliza Hall Institute
    of Medical Research
Monash University
Victoria
Australia

David Horn
London School of Hygiene
    and Tropical Medicine
London, England
UK

K.G. Jayanarayan
Department of Neuroscience
Mayo Clinic
Jacksonville, Florida
USA

Irit Kapeller
Department of Parasitology
The Kuvin Center for the Study
    of Infectious and Tropical Diseases
The Hebrew University-Hadassah
    Medical School
Jerusalem
Israel

Lukasz Kedzierski
Walter and Eliza Hall Institute
    of Medical Research
Monash University
Victoria
Australia

Scott M. Landfear
Department of Molecular
    Microbiology and Immunology
Oregon Health & Science University
Portland, Oregon
USA

Hemanta K. Majumder
Molecular Parasitology Laboratory
Indian Institute of Chemical Biology
Kolkata
India

Chitra Mandal
Immunobiology Division
Indian Institute of Chemical Biology
Kolkata
India

Neta Milman
Department of Parasitology
The Kuvin Center for the Study
    of Infectious and Tropical Diseases
The Hebrew University-Hadassah
    Medical School
Jerusalem
Israel

R.E. Morgan
Division of Medicinal Chemistry
    and Pharmacognosy
College of Pharmacy
The Ohio State University
Columbus, Ohio
USA

Peter J. Myler
Seattle Biomedical Research Institute
and
Departments of Pathobiology
  and Medical Education
  and Biomedical Informatics
Division of Biomedical
  and Health Information
University of Washington
Seattle, Washington
USA

Dotan Sela
Department of Parasitology
The Kuvin Center for the Study
  of Infectious and Tropical Diseases
The Hebrew University-Hadassah
  Medical School
Jerusalem
Israel

Banibrata Sen
Division of Infectious Diseases
Indian Institute of Chemical Biology
Jadavpur, Kolkata
India

Joseph Shlomai
Department of Parasitology
The Kuvin Center for the Study
  of Infectious and Tropical Diseases
The Hebrew University-Hadassah
  Medical School
Jerusalem
Israel

Gaganmeet Singh
Department of Biotechnology
National Institute of Pharmaceutical
  Education and Research
Punjab
India

Alessandro D. Uboldi
Walter and Eliza Hall Institute
  of Medical Research
Monash University
Victoria
Australia

Buddy Ullman
Department of Biochemistry
  and Molecular Biology
Oregon Health & Science University
Portland, Oregon
USA

K.A. Werbovetz
Division of Medicinal Chemistry
  and Pharmacognosy
College of Pharmacy
The Ohio State University
Columbus, Ohio
USA

Nurit Yaffe
Department of Parasitology
The Kuvin Center for the Study
  of Infectious and Tropical Diseases
The Hebrew University-Hadassah
  Medical School
Jerusalem
Israel

Phillip Yates
Department of Biochemistry
  and Molecular Biology
Oregon Health & Science University
Portland, Oregon
USA

Aviad Zick
Department of Parasitology
The Kuvin Center for the Study
  of Infectious and Tropical Diseases
The Hebrew University-Hadassah
  Medical School
Jerusalem
Israel

CONTENTS

ACKNOWLEDGEMENTS

The editor wishes to thank all the eminent scientists who have contributed to this book. He also gratefully acknowledges the support of his students and colleagues in preparing this publication.

CHAPTER 1

Arsenite Resistance in Leishmania and Possible Drug Targets

Gaganmeet Singh, K.G. Jayanarayan and Chinmoy S. Dey*

Abstract

Parasitic infections are of enormous public health importance. Leishmaniasis is currently regarded as the second-most dreaded parasitic disease after malaria (WHO). Visceral leishmaniasis or kala-azar, caused by *Leishmania donovani*, is the most fatal form of leishmaniasis afflicting millions of people worldwide. No vaccination is available against leishmaniasis and fast spreading drug resistance in these parasitic organisms is posing a major medical threat. All these emphasize the need for new drugs and molecular targets along with reappraisal of existing therapeutics. Identification and characterization of cellular targets and answering the problem of drug resistance in *Leishmania* has always been the main thrust of protozoal research worldwide. Model drug resistance phenotypes against drugs, viz. arsenite (an antimony related metal ion, the first line of treatment against leishmaniasis), have been widely used to address and understand mechanism of drug resistance. The present discussion is an attempt to understand the different factors associated with arsenite resistance in *Leishmania*.

Introduction

Leishmaniasis is a group of diseases caused by kinetoplastid protozoan parasite *Leishmania sp*. The causative organism of leishmaniasis are endemic in many parts of world and lead to three major clinical manifestation in human ranging from self-curing cutaneous lesions, noncuring disseminated mucocutaneous to life threatening visceral leishmaniasis. World Health Organization has shown concerns over the enormity of leishmaniasis that has been underrated for long chiefly due to nonreporting of cases in remote areas and the social stigma associated with the deformities.[1] Impact of the leishmaniasis on public health can be judged from the expansion of endemic regions in last one decade, 12 million affected cases spread across 88 countries, 2 million new cases every year and annual mortality rate of more than 60,000 majority of which are children.[1] Reports of coinfection of leishmaniasis with HIV in immuno-compromised host in more than 33 countries and its crossing the barrier of endemic regions has further aggravated the problem.[2]

Control of leishmaniasis continues to be elusive due to the absence of effective vaccines and efficient vector control measures; as a consequence chemotherapy remains the main weapon to combat the disease. Conversely, the lack of range of effective and nontoxic drugs; variation in efficacy as result of intrinsic variation in drug sensitivity; and the emergence of drug resistance limits the arsenal of antileishmanial drugs.[3] It is the developing regions of the world, which are

*Corresponding Author: Chinmoy S. Dey—Department of Biotechnology, National Institute of Pharmaceutical Education and Research (NIPER), Sec. 67, S.A.S. Nagar, Punjab-160 062, India. Email: csdey@niper.ac.in

Drug Targets in Kinetoplastid Parasites, edited by Hemanta K. Majumder.
©2008 Landes Bioscience and Springer Science+Business Media.

predominantly facing the wrath of drug resistance due to unavailable, expensive, newer chemo-therapeutic agents. Steadily progressing chemoresistance to pentavalent antimonials (first line chemotherapy and arsenite related metal ion)[3] and 3-5 fold difference in resistance observed between different *Leishmania* species against other alternative drugs like paramomycin, azoles and miltefosine[4] urge to develop better drugs and identify new drug targets. Thus, unraveling the molecular mechanisms rendering the parasite chemoresistant is important to the discovery of new cellular targets and increasing our arsenal against this parasite.[5,6]

Arsenite Resistance in Leishmania

Long clinical use of arsenicals and related antimony containing drugs in parasitic chemo-therapy has resulted in undefined mechanism of resistance to these drugs in protozoans. Resis-tance to clinical drugs, the major impediment in the treatment of protozoal infection, has always counted on the ease to develop drug resistant in vitro cell lines that has been instrumen-tal in understanding the mechanism of drug resistance.[5,6] There have been number of reports for generation of an in vitro sodium arsenite (oxyanion) resistant cell lines in different *Leishmania sp.* to understand basic molecular mechanisms of drug resistance.[7-11]

Drug resistance phenotype can result due to any of the following possibilities: (i) decrease in drug uptake; (ii) efflux of drug from parasite; (iii) loss of drug activatio; and (iv) alteration of drug targets. Reports available till date suggests that the resistance to the oxyanion, arsenite, in the parasite *Leishmania* is multifactorial and involves events like xenobiotic conjugation and traffic, cytoskleleton phosphorylation, altered expression of different genes and enzyme sys-tems. The following paragraphs detail the current concepts and advancements in this field.

DNA amplification has evolved as a very common means adopted by these parasites against drug pressure.[12-15] Arsenic was found to be potent inducers of gene amplification in *Leishmania*.[7,13,16] In most of the cases the gene amplified in response to drug is present as an extrachro-mosomal circular DNA molecule and code for enzyme or transporter system involved in detoxi-fication process of drug.[7-9,13,17-19] As found previously in methotrexate[13] or tunicamycin[18] resistant strains, duplicated parts of chromosomal region exist as extrachromosomal circular H-circle in arsenite drug variants of *Leishmania*.[7] Direct involvement of 69kb H-circle was observed in resistance to arsenite as variants that revert to wild type after differentiation lack this extra DNA element. Two loci were found to be present in multiple copies in drug resistant variants. One is the H-locus, which harbors a P-glycoprotein related gene, ltpgpA, the first MDR homologue reported in *L. tarentolae*.[20] Second locus is a 50kb linear amplicon of un-known function.[17] ltpgpA and its homologue lmpgpA (reported from *L. major*)[21] was reported to confer resistance to both trivalent arsenite As(III) and trivalent antimony Sb(III). Earlier P-glycoprotein related gene products of ltpgpA and lmpgpA were thought to be main players in mechanism for arsenite resistance. But it was later found that these amplicons could be attributed to low level of resistance to arsenite and antimonite and raise the possibility of the involvement of other mechanisms.[22]

Arsenite resistant *L. tarentolae* has been reported to accumulate less arsenite than parental wild type cell line[23] even when rate of arsenite accumulation was observed to be same in plasma membrane-enriched vesicle prepared from two strains.[24] This suggested the involvement of some other drug resistance mechanism(s) being operative. Radioactive labeling[23] and atomic absorption spectroscopic[25] studies strongly suggested the involvement of an active drug extru-sion system in arsenite resistance. Further the gene disruption studies of ltpgpA in *L. tarentolae* wild type and arsenite resistant strains showed that PgpA is not essential for resistance to oxyanions, although it might be required in the early steps of selection when resistance is being established.[26] Eventually an ATP-coupled pump was identified to be involved in extrusion of metal-thiolates, i.e., thiolate derivatives of As(III).[27] Low rate of transport was observed for free arsenite, but rapid accumulation was observed after reaction with reduced glutathione (GSH), conditions that favor the formation of As(GS)3. These reports suggest a novel mechanism in which pentavalent arsenite and antimony containing compounds are reduced to trivalent species intracellularly, and then these trivalent metal containing compounds conjugate to trypanothione, and are extruded by the As-thiol pump.[28] The rate limiting step in this unique drug extrusion and ultimately drug resistance mechanism was proposed to be the substrate formation i.e., metal-thiol conjugate formation instead of quantity of transporter present on the membrane. But arsenite resistant *L. donovani* has been reported for low level of cross-resistance to structurally unrelated drugs raising the possibility of active export of drug from parasite.[9] Using the human reactive antibody it has been shown to over-express a putative P-glycoprotein[29] with associated enhanced membrane P-type ATPase activity.[11] The above P-glycoprotein was shown to be down regulated by verapamil along with consequent reduction in membrane associated ATPase activity.[11] A recent report has implicated the tryparedoxin peroxidases to be one of the players in arsenite resistance mechanism.[30] It is suggested that the distinct overexpression of cytoplasmic and mitochondrial (cTXNPx and mTXNPx, respectively) genes in arsenite-resistant variants with and without DNA amplification is linked to arsenite selection process.

Potential Drug Targets

Understanding the evolution of resistance by identifying and validating the target molecules involved in resistance can guide in developing strategies to prevent ineffective and often toxic chemotherapy. One very common phenomenon, as already mentioned, observed during the drug resistance is the alteration of drug targets. Our group has been involved in characterizing the microtubules and topoisomerases, which have emerged as the potential molecular targets for many antileishmanial agents directed against arsenite resistant *Leishmania*.

Tubulins

Microtubules are cytoskeletal polymers consisting of repeating α/β-tubulin heterodimers and a variety of minor components known as microtubule-associated proteins which are important for the regulation and distribution of microtubules in the cell.[31] Microtubules play the key role in many vital cellular activities such as mitosis, intracellular vesicle transport, organization and positioning of membranous organelles, and determination of cell shape and motility.

In kinetoplastida, including the parasite *Leishmania*, microtubules are classified in three classes of the flagellar, mitotic and subpellicular microtubules, which are involved in locomotion, cell division and cell shape, respectively. In *Leishmania*, the microtubules have significant differences in their primary amino acid sequence and polymerization properties,[32] which provides parasite specific target for selective chemotherapy.[33,34] The abundant distribution of cytoskeleton in *Leishmania*, the reported affinity of arsenite to microtubules in vitro and high toxicity of arsenite to different *Leishmania* species in its trivalent state,[29] presents the tubulin as one of the ideal targets for arsenite insult. α-tubulin has been successfully correlated in arsenite resistance phenomenon in *L. donovani*.[35] Alteration in posttranslational status of both α- and β-tubulin was also observed in arsenite resistant variants of *L. donovani* with hyper-phosphorylation on serine and threonine residues.[36] Significant differences in the profile of β- and γ-tubulin levels have been reported in resistant variants during various stages of differentiation from promastigotes to amastigotes.[37] Further, changes in the expression of α- and β-tubulin, altered dynamics of microtubule formation and deregulation of the cellular distribution of α- and β-tubulin have been reported to be a part of the response of *L. donovani* promastigotes to arsenite.[38]

Antimicrotubule agents represent a unique class of compounds in so far as their target does not involve nucleic acid synthesis or integrity. Antimicrotubule agents basically exploit the 'dynamic character' the most prominent features of all microtubules (except stable microtubules of cilia and flagella) i.e., the rapid exchange of subunits between polymers and a soluble tubulin pool.[39,40] This dynamic instability is described by four parameters: the rates of 'polymerization' and 'depolymerization', and the frequencies of 'catastrophe' (the transition from polymerization to depolymerization) and 'rescue' (the transition from depolymerization to polymerization).[41] The ability of some compounds to act selectively on one of these parameters decides their usefulness. Selectivity of antimicrotubule agents for nonhuman cells, such as *Leishmania*, has been shown to be due to differences in the binding ability of these compounds to human vs. nonhuman tubulins and the evolution of variations in the primary structure and the emergence of various isotypes of tubulin.[42]

Paclitaxel, a potent inhibitor of cell replication that enhances the polymerization of tubulin into stable bundles of microtubules,[43] has been observed to be two fold more effective in arsenite resistant *L. donovani*.[35] Paclitaxel has been reported in altering expression of different tubulin isoforms in arsenite resistant *L. donovani*.[35,37] With the increase in interest in tubulin as a target site, tubulin inhibiting compounds such as benzimidazoles and the dinitroanilines are being examined for their antiprotozoal effects.[43] The antiparasitic effect of dinitroanilines is thought to originate with the sequence of the gene(s) coding for a tubulin subunit.[43] Anti-leishmanial activity and efficacy of trifluralin, a tubulin-depolymerizing agent, have been established against the arsenite resistant strain of *L. donovani*.[44] Studies have shown that both paclitaxel and trifluralin are effective in limiting parasite growth. Specific alterations in morphology, tubulin polymerization dynamics, post-translational modifications and cellular distribution of the tubulins have been confirmed to be a part of the intracellular anti-microtubule events that occur in arsenite resistant *L. donovani* in response to these agents, ultimately leading to apoptosis like death of the parasite.[44]

Possibility of using a combination of paclitaxel and trifluralin at half of IC_{50} concentrations of each to inhibit both wild type and arsenite resistant *L. donovani* has also been explored. Results obtained have strengthened the hypothesis that the combination of a drug that specifically binds to parasite tubulin, such as trifluralin with paclitaxel results in the exploitation of the hypersensitivity of the resistant strain to paclitaxel along with the species specificity of trifluralin.[44] The combination of trifluralin and paclitaxel at respective half of IC_{50} concentrations in wild type and resistant strain showed an additive effect inhibiting ~55% of parasite growth. By combining trifluralin with paclitaxel, which has considerable toxicity towards mammalian cells,[43] the concentration of paclitaxel required to inhibit (by 50%) the resistant strain could be reduced. Further investigations in vivo are necessary to determine the dependence of these effects on simultaneous or sequential drug addition, dose or duration of treatment.

In the recent years a lot of impetus has been on establishing and exploring the protozoan parasitic programmed cell death (PCD) pathway not only to find out new molecular targets but also to understand the evolution of this phenomenon in these primitive unicellular organisms. The cascade of events taking place in wild type and oxyanion variant *L. donovani* promastigotes upon treatment with antimicrotubule drugs has been reported. Drugs targeting tubulins like sodium arsenite and paclitaxel have been shown to induce PCD in wild type and arsenite resistant strain at their respective IC_{50}.[38,43] Unique features like, PARP-dependent apoptosis in wild type against PARP-independent apoptosis in arsenite variant, was observed. PARP cleavage was not observed on paclitaxel treatment.[31] But in all the cases the regular features of metazoan apoptosis like, circularization of cells, DNA fragmentation with TUNEL positive cells was observed. Trifluralin, another anti-microtubule drug did not induce PCD in *L. donovani*. But at the same time the combination of paclitaxel and trifluralin was able to induce PCD in *L. donovani* promastigotes.[43]

Development of chemotherapeutic strategies based on above antimicrotubule agents against drug resistant *Leishmania* parasites merits further investigations both in vivo and on the amastigote forms of the parasite since variations in the basal levels of tubulins between the promastigotes and amastigotes has already been reported.[35,37] Havens et al[45] have reported many-fold increased susceptibility of amastigotes to taxol as compared to promastigotes of *Leishmania*. Further extension of these investigations to determine the effects of the antimicrotubule agents on strains resistant to other clinically used drugs may provide useful information.

Topoisomerases

DNA topoisomerases are the enzymes that remove torsional stress in DNA by introducing transient protein-bridged DNA breaks on one (type I) or both DNA strands (type II).[47] By regulating DNA topology during replication, transcription, recombination, repair of DNA and also for chromosome (de)condensation and sister chromatid segregation, they play an essential role in the maintenance of genetic material integrity.[47] Topoisomerases has been identified as one of the major targets in cancer and bacterial chemotherapy.[48,49] Protozoan parasites has unique requirement of topoisomerases, especially topoisomerase II owing to presence of complex intercatenated network of thousands of minicircle and few dozens of maxicircle in kinetoplastid's mitochondria (kDNA).[50] Apart form DNA amplification of chromosomal regions, alteration in the nucleic acids of mitochondrial DNA i.e., kinetoplast DNA has been reported in *Leishmania* on arsenite pressure.[51-53] Arsenite has been reported to induce changes in sequence of maxicircle DNA,[51] dominance and selection of particular class of minicircle[52] and changes in structural organization of kinetoplast in *L. mexicana*.[53] Though the physiological consequence of these alterations is still not clear but possibility of topoisomerase II to be victim of arsenite insult cannot be ruled out, as it is involved in maintaining the integrity of nucleic acid material. Drug specific up- or down-regulation of topoisomerase II is an established fact in mammalian cells. Changes in expression of topoisomerase II have been associated with appearance of antitumor drug resistance.[54] Topoisomerase II has been reported to be over-expressed and observed to show increase in activity in arsenite resistant *L. donovani*.[55,56] These raise the possibility of topoisomerase II being the part of multifactorial effect of arsenite. More than 3-fold increase in topoisomerase II specific activity and nearly 2-fold increase in topoisomerase II expression level were observed in crude nuclear extract of arsenite resistant variants as compared to wild type.[56]

Immense topological problems posed by kDNA[57] presents potential site for topoisomerase II activity. The essentiality of topoisomerases for maintaining genetic integrity and vitality for cell survival lead them to be targeted by several clinically important antibacterial and anticancerous drugs like fluroquinolones, etoposide, camptothecins and novobiocin.[58] Effect of novobiocin, a coumarin class of antibiotic and known catalytic inhibitor of DNA gyrase, has been investigated on topoisomerase II of the arsenite resistant and parent cell line in *L. donovani*. It has been shown that novobiocin differentially inhibit the topoisomerase II activity in two strains.[56] Significant differences have been observed between the responses of wild type and resistant promastigotes to novobiocin treatment with arsenite resistant strain being cross-resistant to novobiocin.[56] Requirement of higher drug concentrations, to inhibit topoisomerase II activity and cell growth in arsenite resistant *L. donovani* variant, again advocate the presence of unique or excess topoisomerase II activity in resistant phenotype. Higher inhibition observed in arsenite resistant falls well in line with earlier reports from our group that showed arsenite resistant strain to be cross-resistant to pentamidine.[11] Studies on topoisomerase II targeted drug interaction will aid in further development of chemotherapeutic strategies based on anti-topoisomerase II agents against drug resistant *Leishmania* parasites. To the best of our knowledge, for the first time the topoisomerase II has been implicated in arsenite resistance. Drug targeting topoisomerase II like topoisomerase II inhibitor novobiocin have also been observed to induce PCD in wild type and arsenite resistant *L. donovani* at different concentrations.[56] The metazoan features of apoptosis like externalization of phosphatidyl serine, release of cytochrome C, activation of cellular proteases, DNA laddering and oligonucleosomal fragmentation was observed as a part of anti-topoisomerase II agents induced apoptosis in *L. donovani*.

Conclusion

In conclusion, above studies provide valuable insights into multifactorial nature of oxyanion (arsenite) resistance in *Leishmania*. Several metabolic pathways and membrane transporters are implicated in the resistance phenotype. A recent report utilizing transcript profiling of arsenite and antimony resistant mutants with microarrays pinpointed altered expression of number of genes in mutants. The genes include the ABC transporter PGPA, the glutathione biosynthesis genes gamma-glutamylcysteine synthetase (GSH1) and the glutathione synthetase (GSH2) which further strengthened the belief that drug resistance can be a complex phenomenon in the protozoan parasite *Leishmania*.[60] Apoptosis in protozoan parasites represent a philanthropic device adopted by these parasites to furnish the best population to be transmitted to the next host and unraveling the pathway in parasite presents us the potentiality of finding newer and specific targets. The advent of proteomics has opened a new chapter in understanding not only the parasitic biology but the actual holistic effect of drug resistance.[61] Taken together, the above findings provide valuable information in further development of chemotherapeutic strategies based on overexpressing enzyme systems (glutathione/typanothione metabolism; topoisomerase II) and antimicrotubule agents against drug resistant *Leishmania* parasites. The success of these prospective antileishmanial agents on amastigotes form and in vivo await further investigations.

Acknowledgements

GS and KGJ were supported by Junior/Senior Research Fellowship from Council for Scientific and Industrial Research, New Delhi, Govt. of India. The work was funded by a grant from Council for Scientific and Industrial Research, New Delhi, Govt. of India.

References

1. http://www.who.int/leishmaniasis.
2. Russo R, Laguna F, Lopez-Velez R et al. Visceral leishmaniasis in those infected with HIV: Clinical aspects and other opportunistic infections. Ann Trop Med Parasitol 2003; 97:99-105.
3. Croft SL, Coombs GH. Leishmaniasis—current chemotherapy and recent advances in the search for novel drugs. Trends Parasitol 2003; 19:502-508.
4. Croft SL. Monitoring drug resistance in leishmaniasis. Trop Med Int Health 2001; 6:899-905.
5. Oullette M, Drummelsmith J, Papadopoulou B. Leishmaniasis: Drugs in clinic, resistance and new developments. Drug Resist Updat 2004; 7:257-66.
6. Ponte-Sucre A. Physiological consequences of drug resistance in Leishmania and their relevance for chemotherapy. Kinetoplastid Biol Dis 2003; 2:14.
7. Detke S, Katakura K, Chang KP. DNA amplification in arsenite resistant Leishmania. Exp Cell Res 1989; 180:161-170.
8. Ouellette M, Hettema E, Wust D et al. Direct and inverted DNA repeats associated with P-glycoprotein gene amplification in drug resistant Leishmania. EMBO J 1991; 10:1009-1016.
9. Callahan HL, Beverley SM. Heavy metal resistance: A new role for P-glycoproteins in Leishmania. J Biol Chem 1991; 266:18427-30.
10. Sho-Tone L, Tarn C, Wang CY. Characterization of sequence changes in kinetoplast DNA maxicircles of drug-resistant Leishmania. Mol Biochem Parasitol 1992; 56:197-208.
11. Prasad V, Kaur J, Dey CS. Arsenite-resistant Leishmania donovani promastigotes express an enhanced membrane P-type adenosine triphosphatase activity that is sensitive to verapamil treatment. Parasitol Res 2000; 86:661-664.
12. Beverley SM. Gene amplification in Leishmania. Annu Rev Microbiol 1991; 45:417-44.
13. Ouellette M, Borst P. Drug resistance and P-glycoprotein gene amplification in the protozoan parasite Leishmania. Res Microbiol 1991; 142:737-746.
14. Segovia M. Leishmania gene amplification: A mechanism of drug resistance. Ann Trop Med Parasitol 1994; 88:123-30.
15. Schallig HD, Oskam L. Molecular biological applications in the diagnosis and control of leishmaniasis and parasite identification. Trop Med Int Health 2002; 7:641-51.
16. Ouellette M, Papadopoulou B. Mechanism of drug resistance in Leishmania. Parasitol Today 1993; 9:150-153.
17. Grondin K, Papadopoulou B, Ouellette M. Homologous recombination between direct repeats sequences yields P-glycoprotein containing circular amplicons in arsenite resistant Leishmania. Nucleic Acids Res 1993; 21:1895-1901.
18. Detke S, Chaudhuri G, Kink JA et al. DNA amplification in tunicamycin-resistant Leishmania mexicana. Multicopies of a single 63-kilobase supercoiled molecule and their expression. J Biol Chem 1988; 263:3418-3424.

19. Beverley SM, Coderre JA, Santi DV et al. Unstable DNA amplifications in methotrexate-resistant Leishmania consist of extrachromosomal circles which relocalize during stabilization. Cell 1984; 38:431-439.
20. Ouellette M, Fase-Fowler F, Borst P. The amplified H circle of methotrexate resistant Leishmania contains a novel P-glycoprotein gene. EMBO J 1990; 9:1027-1033.
21. Callahan HL, Beverley SM. Heavy metal resistance: A new role for P-glycoproteins in Leishmania. J Biol Chem 1991; 266:18427-30.
22. Papadopoulou B, Roy G, Dey S et al. Contribution of the Leishmania P-glycoprotein-related gene ltpgpA to oxyanion resistance. J Biol Chem 1994; 269:11980-6.
23. Dey S, Papadopoulou B, Haimeur A et al. High level resistance in Leishmania tarentolae is mediated by an active extrusion system. Mol Biochem Parasitol 1994; 67:49-57.
24. Grondin K, Haimeur A, Mukhopadhyay R et al. Coamplification of the gamma-glutamylcysteine synthetase gene gsh1 and of the ABC transporter gene pgpA in arsenite-resistant Leishmania tarentolae. EMBO J 1997; 6:3057-3065.
25. Singh AK, Liu HY, Lee ST. Atomic absorption spectrophotometric measurement of intracellular arsenite in arsenite-resistant Leishmania. Mol Biochem Parasitol 1994; 66:161-164.
26. Papadopoulou B, Roy G, Dey S et al. Gene disruption of the P-glycoprotein related gene pgpa of Leishmania tarentolae. Biochem Biophys Res Commun 1996; 224:772-778.
27. Dey S, Ouellette M, Lightbody J et al. An ATP-dependent As(III)-glutathione transport system in membrane vesicles of Leishmania tarentolae. Proc Natl Acad Sci USA 1996; 93:2192-2197.
28. Mukhopadhyay R, Dey S, Xu N et al. Trypanothione overproduction and resistance to antimonials and arsenicals in Leishmania. Proc Natl Acad Sci USA 1996; 93:10383-10387.
29. Kaur J, Dey CS. Putative P-glycoprotein expression in arsenite-resistant Leishmania donovani down-regulated by verapamil. Biochem Biophys Res Commun 2000; 271:615-619.
30. Lin YC, Hsu JY, Chiang SC et al. Distinct overexpression of cytosolic and mitochondrial tryparedoxin peroxidases results in preferential detoxification of different oxidants in arsenite-resistant Leishmania amazonensis with and without DNA amplification. Mol Biochem Parasitol 2005; 142:66-75.
31. Jayanarayan KG, Dey CS. Microtubules: Dynamics, drug interaction and drug resistance in Leishmania. J Clin Pharm Therapeut 2002; 27:313-320.
32. Werbovetz KA, Brendle JJ, Sackett DL. Purification, characterization, and drug susceptibility of tubulin from Leishmania. Mol Biochem Parasitol 1999; 98:53-65.
33. Croft SL. Recent developments in the chemotherapy of leishmaniasis. Trends Pharmacol Sci 1988; 9:376-381.
34. Seeback T, Hemphill A, Lawson D. The Cytoskeleton of trypanosomes. Parasitol Today 1990; 6:49-52.
35. Prasad V, Kumar SS, Dey CS. Resistance to arsenite modulates levels of alpha-tubulin and sensitivity to paclitaxel in Leishmania donovani. Parasitol Res 2000; 86:838-842.
36. Prasad V, Dey CS. Tubulin is hyperphosphorylated on serine and tyrosine residues in arsenite-resistant Leishmania donovani promastigotes. Parasitol Res 2000; 86:876-880.
37. Jayanarayan KG, Dey CS. Resistance to arsenite modulates expression of beta- and gamma-tubulin and sensitivity to paclitaxel during differentiation of Leishmania donovani. Parasitol Res 2002; 88:754-759.
38. Jayanarayan KG, Dey CS. Altered expression, polymerisation and cellular distribution of alpha-/beta-tubulins and apoptosis-like cell death in arsenite resistant Leishmania donovani promastigotes. Int J Parasitol 2004; 34:915-925.
39. Dumontet C. Mechanism of action and resistance to tubulin binding agents. Expert Opinion Inves Drugs 2000; 9:779-788.
40. Rodionov V, Nadezhdina E, Boriy G. Centrosomal control of microtubule dynamics. Proc Natl Acad Sci USA 1999; 96:115-120.
41. Margolis RL, Wilson L. Microtubule treadmilling: What goes around comes around. Bioessays 1998; 20:830-836.
42. Little M, Seehaus T. Comparative analyses of tubulin sequences. Comp Biochem Physiol 1988; 90:655-670.
43. Ojima I. A new paclitaxel photoaffinity analog with a 3-(4-benzoylphenyl)propanoyl probe for characterization of drug-binding sites on tubulin and P-glycoprotein. J Med Chem 1995; 38:3891-3894.
44. Jayanarayan KG, Dey CS. Altered tubulin dynamics, localization and post-translational modifications in sodium arsenite resistant Leishmania donovani in response to paclitaxel, trifluralin and a combination of both and induction of apoptosis-like cell death. Parasitology 2005; 131:215-230.

45. Werbovetz KA. Tubulin as an antiprotozoal drug target. Mini Rev Med Chem 2002; 2:519-529.
46. Havens CG, Bryant N, Asher L et al. Cellular effects of leishmanial tubulin inhibitors on Leishmania donovani. Mol Biochem Parasitol 2000; 110:223-36.
47. Wang JC. Cellular roles of DNA topoisomerases: A molecular prospective. Nat Rev Mol Cell Biol 2002; 3:430-440.
48. Schneider E, Hsiang YH, Liu LF. DNA topoisomerases as antitumor drug targets. Adv Pharmacol 1991; 21:149-183.
49. Heisig P. Inhibitors of bacterial topoisomerases: Mechanisms of action and resistance and clinical aspects. Planta Med 2001; 67:3-12.
50. Shlomai J. The structure and replication of kinetoplast DNA. Curr Mol Med 2004; 4:623-647.
51. Lee ST, Tarn C, Wang CY. Characterization of sequence changes in kinetoplast DNA maxicircles of drug-resistant Leishmania. Mol Biochem Parasitol 1992; 56:197-207.
52. Lee ST, Tarn C, Chang KP. Characterization of the switch of kinetoplast DNA minicircle dominance during development and reversion of drug resistance in Leishmania. Mol Biochem Parasitol 1993; 58:187-203.
53. Lee ST, Liu HY, Lee SP et al. Selection for arsenite resistance causes reversible changes in minicircle composition and kinetoplast organization in Leishmania mexicana. Mol Cell Biol 1994; 14:587-596.
54. Pu QQ, Bezwoda WR. Alkylator resistance in human B lymphoid cell lines: (2). Increased levels of topoisomerase II expression and function in a melphalan-resistant B-CLL cell line. Anticancer Res 2000; 20:2569-22578.
55. Jayanarayan KG, Dey CS. Overexpression and increased DNA topoisomerase II-like enzyme activity in arsenite resistant Leishmania donovani. Microbiol Res 2003; 158:55-58.
56. Singh G, Jayanarayan KG, Dey CS. Novobiocin induces apoptosis-like cell death in topoisomerase II over-expressing arsenite resistant Leishmania donovani. Mol Biochem Parasitol 2005; 141:57-69.
57. Shapiro TA, Englund PT. The structure and replication of kinetoplast DNA. Annu Rev Microbiol 1995; 49:117-143.
58. Larsen AK, Escargueil AE, Skladanowski A. Catalytic topoisomerase II inhibitors in cancer therapy. Pharmacol Ther 2003; 99:167-181.
59. Verma NK, Dey CS. Possible mechanism of miltefosine-mediated death of Leishmania donovani. Antimicrob Agents Chemother 2004; 48:3010-3015.
60. Guimond C, Trudel N, Brochu C et al. Modulation of gene expression in Leishmania drug resistant mutants as determined by targeted DNA microarrays. Nucleic Acids Res 2003; 31:5886-5896.
61. Drummelsmith J, Brochu V, Girard I et al. Proteome mapping of the protozoan parasite Leishmania and application to the study of drug targets and resistance mechanisms. Mol Cell Proteomics 2003; 2:146-155.

CHAPTER 2

Unique Characteristics of the Kinetoplast DNA Replication Machinery Provide Potential Drug Targets in Trypanosomatids

Dotan Sela, Neta Milman, Irit Kapeller, Aviad Zick, Rachel Bezalel, Nurit Yaffe and Joseph Shlomai*

Reevaluating the Kinetoplast as a Potential Target for Anti-Trypanosomal Drugs

Kinetoplast DNA (kDNA) is a remarkable DNA structure found in the single mitohondrion of flagellated protozoa of the order Kinetoplastida. In various parasitic species of the family Trypanosomatidae, it consists of 5,000-10,000 duplex DNA minicircles (0.5-10 kb) and 25-50 maxicircles (20-40 kb), which are linked topologically into a two dimensional DNA network. Maxicircles encode for typical mitochondrial proteins and ribosomal RNA, whereas minicircles encode for guide RNA (gRNA) molecules that function in the editing of maxicircles' mRNA transcripts. The replication of kDNA includes the duplication of free detached minicircles and catenated maxicircles, and the generation of two progeny kDNA networks. It is catalyzed by an enzymatic machinery, consisting of kDNA replication proteins that are located at defined sites flanking the kDNA disk in the mitochondrial matrix (for recent reviews on kDNA see refs. 1-8).

The unusual structural features of kDNA and its mode of replication, make this system an attractive target for anti-trypanosomal and anti-leishmanial drugs. However, in evaluating the potential promise held in the development of drugs against mitochondrial targets in trypanosomatids, one has to consider the observations that dyskinetoplastic (Dk) bloodstream forms of trypanosomes survive and retain their infectivity, despite the substantial loss of their mitochondrial genome (recently reviewed in ref. 9). Survival of Dk strains has led to the notion that kDNA and mitochondrial functions are dispensable for certain stages of the life cycle of trypanosomatids. This view has been challenged by Schnaufer et al,[10] who demonstrated that knock-down of RNA ligase in bloodstream forms of *Trypanosoma brucei*, was lethal to the parasite. Furthermore, in a recent report[11] they have demonstrated that silencing the expression of the α-subunit of mitochondrial F_1-ATP synthase complex, was lethal to bloodstream stage *Trypanosoma brucei*, as well as to the dyskinetoplastic species *Trypanaosoma evansi*. Schnaufer et al have suggested[9] that the lethality resulting from the loss of kDNA, or the lack of expression of its encoded genes, could be due to several possible reasons. One possibility is that several kDNA genes may have an essential role in the bloodstream stage of the parasite. In accord with this notion is the case of silencing the F_1-ATP synthase, where the lethal effect has

*Corresponding Author: Joseph Shlomai—Department of Parasitology, The Kuvin Center for the Study of Infectious and Tropical Diseases, The Hebrew University- Hadassah Medical School, Jerusalem 91120, Israel. Email: Shlomai@cc.huji.ac.il

Drug Targets in Kinetoplastid Parasites, edited by Hemanta K. Majumder.
©2008 Landes Bioscience and Springer Science+Business Media.

apparently resulted from the collapse of the mitochondrial membrane potential.[11] As mitochondrial division and cytokinesis are highly coordinated, it is also possible that cell lethality results from the requirement for kDNA in order to conduct a normal process of cell division, which is mediated through specific checkpoints linking the cell cycle to kDNA segregation.[12-14] It is also possible that the lack of kDNA results in triggering of a series of events that lead to programmed cell death.[15,16]

These observations, suggesting that mitochondrial functions are not dispensable in bloodstream parasites, raise interest in mitochondrial targets for the development of drugs against pathogenic trypanosomatids. The following chapter describes recent advances in the study of kDNA replication, emphasizing the unique features of this system.

The kDNA Network and Its Monomeric Components

One of the most extensively studied kDNA networks is that of the species *Crithidia fasciculata*. The kDNA network (approximately 10 by 15 μm in dimensions[17]) is condensed in the mitochondrial matrix into a disk-like structure of about 1 by 0.35 μm.[18] Several histone-like proteins are involved in the structural organization of the condensed network.[19-25] The kinetoplast was shown by biochemical and molecular studies to be physically attached to the basal body.[3,14,26,27]

The *C. fasciculata* kDNA network consists of ~5,000 minicircles of 2.5 kb and ~25 maxicircles of 37 kb. Minicircles in the network are relaxed and singly interlocked to each other[28,29] forming a two dimensional DNA network. Maxicircles form independent topological catenanes,[30] that are threaded into the minicircles network and are embedded in different patterns within kDNA networks in the various trypanosomatid species, to form 'network within a network'.[30,31]

Maxicircles, the trypanosomal equivalent of mitochondrial genomes in other eukaryotic cells, are approximately identical in size within a given species, but vary in size (19-39 kb) in different trypanosomatids.[32,33] They consist of a conserved coding region and a nontranscribed variable region. Maxicircles genome encodes typical mitochondrial products, such as ribosomal RNA and protein subunits of the respiratory chain,[34] but not for mitochondrial tRNAs, which are encoded by nuclear genes.[35,36] Their transcripts undergo a remarkable process of post-transcriptional editing that includes insertions and deletions of uridine residues, to create functional ORFs (recently reviewed in refs. 7,8,37).

Whereas it has long been known that maxicircles' genome contains typical mitochondrial genes, the function of kDNA minicircles, the major constituent of the network, has remained a puzzle for many years. Currently, their only known genetic function is to encode guide RNA (gRNA) molecules that provide the specificity for RNA editing of maxicircles' transcripts.[38] Studies, suggesting that kDNA minicircles may encode for other RNA and protein products in various trypanosomatids, have also been reported.[39-42] Minicircles within the network of a given trypanosomatid species are virtually identical in size, but are heterogeneous in their nucleotide sequence, in an apparent correlation with the extent of RNA editing in the various trypanosomatid species.[33,43-45] Despite this substantial heterogeneity, they all contain conserved regions of 100-200 bp, whose location and copy number vary in different species.[46-52] These regions contain a common sequence motif[46] that consists of three short conserved sequence blocks (CSBs) that are present in the same order and spacing in all species studied. A 12-mer sequence (CSB-3), known as the universal minicircle sequence (UMS), and a 10-mer sequence (CSB-1), were proposed to contain the replication initiation sites for the minicircle light (L) and heavy (H) strands, respectively.[53-59] Minicircles in most trypanosomatid species contain an additional common structural motif of a region forming a local bend in the DNA double helix,[60-62] whose biological function is yet unknown.

Unique Characteristics of the kDNA Replication System

The unusual topology of the kDNA network and its unique mode of replication, division and segregation, pose several major challenges to the trypanosomatid cell. These are addressed by a replication machinery that carries out a replication scheme with no precedent in any other replication system studied. Some of the individual kDNA replication proteins, such as DNA polymerases, ligases and topoisomerases are similar in function and structure, to various degrees, to the respective enzymes in other replication systems. Others, such as the proposed minicircle origin binding protein UMSBP (see below), are unique in structure and function to the kDNA replication machinery. However, the unique features of kDNA and of the process of its replication, provide potential targets for the development of anti-trypanosomal and anti-leishmanial drugs. The general outlines of the kDNA replication scheme and several unique features of the kDNA replication machinery are discussed in the following paragraphs.

Unlike the replication of mammalian mitochondrial DNA that takes place throughout the entire cell cycle (for review see refs. 63-66), replication of kDNA networks occurs during a discrete S-phase of the cell cycle.[67] Kinetoplast S phase initiates immediately before that of the nuclear S phase, but it is considerably shorter and the kinetoplast segregation is completed before the onset of mitosis.[67-70] kDNA replication includes the duplication of minicircles and maxicircles and the division of the replicated network into two progeny networks that subsequently segregate into the two daughter cells.

A model that provided the basic concepts for understanding the replication of kDNA networks[71,72] had been proposed by Paul Englund almost three decades ago (reviewed in ref. 73) and has since been updated and refined. According to this model, minicircles are not replicated while attached to the network. Instead, covalently closed minicircles are released from the network, prior to their replication, through the action of a type II DNA topoisomerase, and replicate as free DNA circles.[72] The resulting progeny minicircles, which are nicked and gapped, reattach to the network, by the action of another DNA topoisomerase II.[74] Following topological remodeling of the network (see below) and the repair of gaps and nicks in the newly-replicated minicircles, the network splits and subsequently segregates during cell division into the two daughter cells.

kDNA replication proteins were localized to defined sites in the mitochondrial matrix, flanking the kDNA disk. Several of these proteins are clustered at these sites during S-phase, in correlation with the progress in the cell cycle and the process of kDNA replication.[75-78] Fluorescence microscopy studies have localized replication proteins to three distinct regions within the mitochondrial matrix (Fig. 1): (i) at the kineto-flagellar zone (KFZ), which is located between the kDNA disk and the flagellar basal body; (ii) at two antipodal sites flanking the kDNA disk; and (iii) throughout the entire network. It is speculated that proteins that are clustered at overlapping location in the mitochondrial matrix are likely to interact with each other during minicircles replication, to form functional complexes that catalyze related replicative activities. KFZ has been proposed as the site where replication of free minicircles occurs (Fig. 1). It contains (i) the universal minicircle sequence binding protein (UMSBP)[76] (Fig. 2), proposed to function as the origin binding protein;[76,79-84] (ii) DNA primase, that can catalyze the synthesis of RNA primers, is localized close at the two faces the kDNA disk;[78,85] (iii) DNA polymerases Pol IB and Pol IC, shown by RNAi analysis to be required for kDNA replication;[86] and (iv) the kinetoplast associated protein 1 (KAP1), a histone-like protein, which surrounds the kDNA disk, overlapping with DNA primase.[85] Historically, replication proteins clusters were first detected at two sites flanking the kDNA disk (Fig. 1), 180° apart on its circumference.[74] The partial repair of newly-replicated minicircles, as well as their reattachment to the network, occurs at these sites. These sites contain (i) DNA topoisomerase II[74] that catalyzes the topological interconversions of free minicircle and catenane networks[87] and has recently been shown, by RNAi analysis, to be essential for the post-replication reattachment of minicircles;[88] (ii) DNA polymerase β[18] (Fig. 2A), whose catalytic properties, including its dRP-lyase activity,[89-92] suggest a function in the gap-filling of newly replicated minicircles; (iii) a structure-specific endonuclease 1 (SSE1),[77] whose proposed function in primer-excision is supported by its catalytic properties,[92,93] as well as by a recent RNAi analysis;[94] and (iv) the recently discovered DNA ligase kβ,[95,96] whose involvement in the sealing of nicks in the

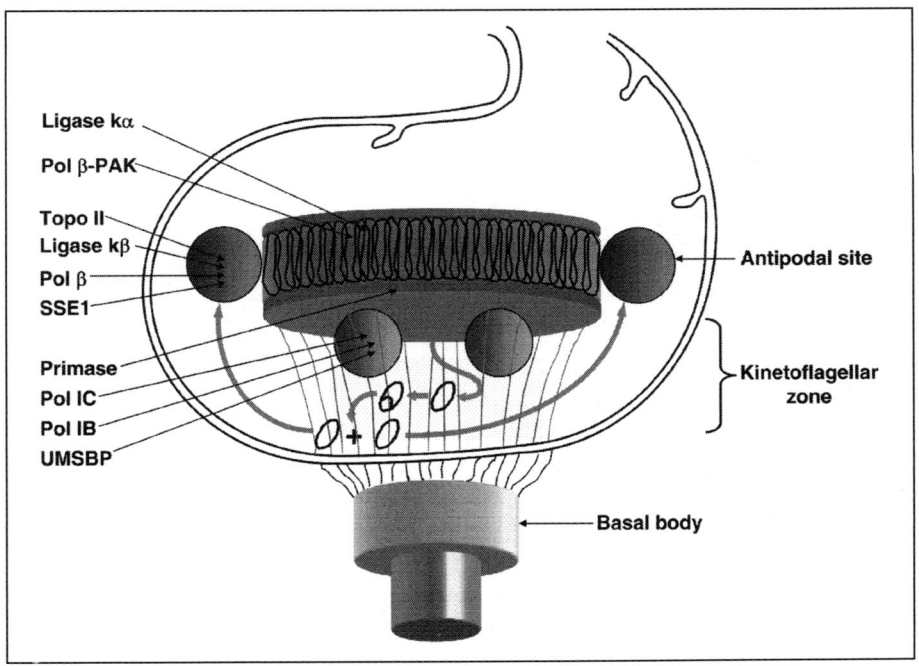

Figure 1. Intramitochondrial location of kDNA replication proteins, and kDNA-replication model. The scheme is based on studies of both *Crithidia fasciculata* and *Trypanosoma brucei*. The kDNA disk, organized with minicircles stretched parallel to its axis, is surrounded by replication proteins. Covalently closed minicircles are released from the network into the KFZ, in which they initiate replication as θ structures [this process probably involves UMSBP, primase, DNA polymerases (Pols) IB and IC, and other proteins]. The progeny free minicircles then migrate to the antipodal sites at which the next stages of replication occur (primer removal by SSE1, gap filling by DNA polymerase β and the sealing of most of the nicks by DNA ligase kβ). The minicircles (still containing at least one nick or gap) are then linked to the network periphery by topoisomerase II (Topo II). DNA polymerase β-PAK and DNA ligase kα are probably involved in the repair of the remaining minicircle gaps when replication is completed. The figure shows the filament system linking the kDNA to the flagellar basal body. Reprinted from: Liu B et al. Fellowship of the rings: The replication of kinetoplast DNA. Trends Parasitol 2005; 21(8):363-9; ©2005 with permission from Elsevier.[4]

newly-replicated minicircles was supported by its coimmunoprecipitation with the mitochondrial DNA polymerase β.[96] Finally, the proteins dispersed throughout the entire kDNA disk, are enzymes involved in the final repair of gapped and nicked progeny minicircles (Fig. 1), and histone-like proteins (KAPs) that are most probably involved in the condensation of the network in the mitochondrial matrix. These include (i) DNA polymerase β-pak,[97] which was suggested to function during the late stage of gap-filling of the reattached minicircles;[97] (ii) the recently discovered DNA ligase kα, whose essential role in the repair of reattached kDNA minicircles was demonstrated recently by RNAi analysis;[96] and (iii) The histone-like proteins KAP2, KAP3 and KAP4.[24]

Another unique feature of the kDNA replication machinery is the mechanisms it utilizes to overcome the major topological challenges in the course of the network replication. Forming a giant topological catenane, which consists of several thousands DNA circles and yet remains confined to a defined space within the mitochondrial matrix, kDNA has to go through dynamic changes in the network topology, termed "remodeling" of the network.[28,98] Prior to replication, each minicircle in the network is interlocked to the average of three other minicircles, yielding a valence of 3. At the end of S-phase, the replicated network contains twice the number of minicircles in the same surface area. As a result, the density of the network increases and its valence is now 6. During G2 phase a remarkable process of topological remodeling of the network occurs, in which the network size increases and its topological valence returns to its prereplication value of 3. This process is followed by the completion of the final steps in the gap-filling and sealing of the topologically linked newly-replicated minicircles.[17]

The subsequent stage of scission of the covalently-sealed, double-size network, demonstrates another remarkable aspect of the kDNA replication machinery. Based on the catenane nature of the network it is presumed that a type II DNA topoisomerase is involved in this process. Since division of the double-size network is almost symmetrical, yielding two daughter networks of approximately the same size,[31,99-101] scission of the network has to be a highly precise process. Performance of such a highly accurate scission has to be tightly controlled by a mechanism that directs the action of the operating topoisomerase to unlink the correct pairs of interlocked minicircles along the virtual line that divides the network into two equal daughter networks. At present, neither the mechanism used for the scission, nor the mode of its regulation is known.

Finally, the segregation of the divided network during cytokinesis is yet another unique feature of the kDNA replication machinery. The molecular connections between the kinetoplast and the basal body has been demonstrated by biochemical and molecular studies, as well as electron microscopy.[14,102,103] It was found that segregation of the basal body drives the separation of the replicated kDNA progeny network, through a microtubules-mediated process,[14,104] resulting in their segregation into the two daughter cells. Segregation of the network was found to be highly coordinated with the process of cytokinesis.[12-14]

Replication of Free kDNA Minicircles and Catenated Maxicircles

Early studies have suggested that replication of the minicircle light (L) strand is continuous, while that of its heavy (H) strand is discontinuous and proceeds through the synthesis of short Okazaki fragments. The conserved sequences at CSB-3 (UMS) and CSB-1 were implicated in minicircle replication initiation, as the functional replication origins for the synthesis of the L and H strands, respectively. Replication of free minicircles initiates by the synthesis of an RNA primer at the conserved UMS site on the H-strand template. The primer is elongated continuously and unidirectionally by a replicative DNA polymerase, displacing the parental L-strand. Subsequently, the discontinuous synthesis of the H-strand is initiated, at the CSB-1 region, and proceeds unidirectionally, using the displaced parental L-strand as a template.[53-59] The mechanism used for priming the synthesis of the minicircle H-strand is yet unknown.

Advances over the past twenty years in the characterization of kDNA replication proteins, their intramitochondrial localization and their role in kDNA replication, have shaped our current view of the process of minicircle replication. According to the refined replication model (Fig. 1), minicircle replication begins by the vectorial release of covalently sealed prereplicated minicircles to the KFZ.[105] This region, which accommodates UMSBP (Fig. 2) and DNA primase, as well as the DNA polymerases Pol IB and Pol IC (Fig. 1), also contains minicircle replication intermediates during S phase.[105] It has been suggested that assembly of a minicircle replication-initiation complex takes place at the KFZ, triggering the priming of the minicircle's leading (L) strand synthesis, and the assembly of the replication fork (reviewed in refs. 3,4). It has been further speculated that synthesis of the minicircle leading and lagging strands, as well as segregation of the daughter minicircles, occur in the KFZ, and that the progeny minicircles migrate from the KFZ to the antipodal reattachment sites. Free minicircle replication intermediates have been detected at these two sites during S-phase (Fig. 2B).[4,18,78] The mechanism that controls the migration of the newly replicated kDNA minicircles to the antipodal sites is yet unknown. While at these sites, prior to their catenation onto the network by the type II DNA topoisomerase,[74,87] the newly replicated minicircles are partially repaired, by the excision of the

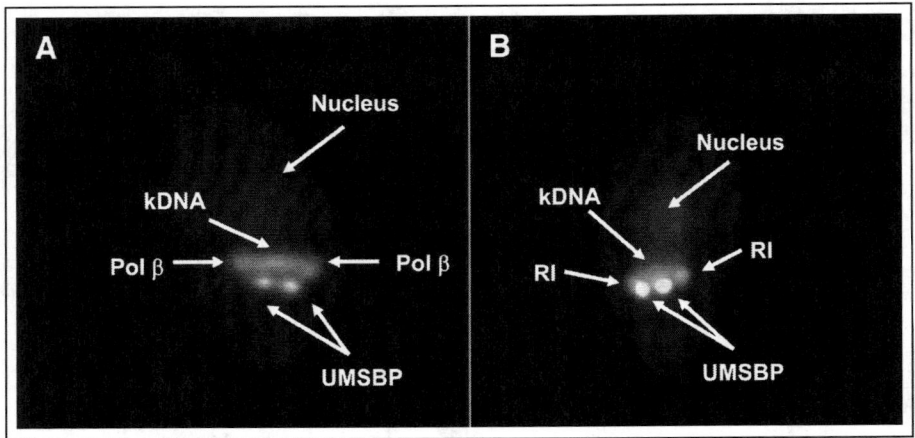

Figure 2. The intra-mitochondrial localization of UMSBP, DNA polymerase β and centers of minicircles replication intermediates (RI). Fluorescence microscopy, demonstrating the localization of replication proteins and minicircle replication intermediates (RI) in the mitochondrial matrix, surrounding the kDNA disk. A) Localization of DNA polymerase β (pol β) and UMSBP: An overlay presentation of DAPI staining (blue) of the kDNA disk, and immunostaining of pol β (red), showing its localization at the two antipodal sites, and of UMSBP (green), at the kineto-flagellar zone (KFZ). B) Antipodal localization of minicircle replication intermediates (RI): An overlay presentation of DAPI staining (blue), Alexa-dUTP fluorescence (red), and UMSBP (green). Minicircle replication intermediates (RI), which are gapped, were selectively labeled in situ by incorporation Alexa-dUTP using terminal deoxynucleotididyl transferase.[76] Reproduced from: Abu-Elneel K et al. Intramitochondrial localization of universal minicircle sequence-binding protein, a trypanosomatid protein that binds kinetoplast minicircle replication origins. J Cell Biol 2001; 153(4): 725-34; by copyright permission of The Rockefeler University Press.[76] A color version of this figure is available online at www.Eurekah.com.

remaining primers, filling of the gaps and sealing of nicks, through the action of SSE1,[77,93] DNA polymerase β (Fig. 2A),[18,90] and ligase kβ.[95,96] Complete repair of discontinuities in the newly replicated minicircles, a process that is carried out in other replication systems during DNA synthesis, is delayed here until all minicircles have been duplicated and reattached to the network. It has been suggested that the final repair of gaps and nicks in the minicircle, which precedes the division of the network, involves the action of DNA polymerase β-PAK[97] and ligase kα.[96] The reason for the delayed repair of newly replicated minicircles is yet unknown. It has been proposed that the presence of one or more nicks and gaps in the reattached minicircles may serve in a 'book-keeping' mechanism, marking replicated minicircles to insure the replication of each minicircle molecule only once per generation.[106,107]

Much less is known about the replication of kDNA maxicircles, which occurs during S-phase, concurrently with the replication of minicircles. Unlike minicircles, that replicate as free detached DNA circles, maxicircles replicate while attached to the network.[108,109] Maxicircles replication initiates from a replication origin, located in their noncoding (variable) region and proceeds unidirectionally through theta (θ) structure intermediates.[108] Involvement of RNA polymerase in maxicircles replication was also reported.[110]

Regulation of kDNA Replication

Considerable progress has been made in recent years in our understanding of the enzymatic machinery that catalyzes the assembly of kDNA networks in trypanosomatids. Nevertheless, our understanding of the mechanisms that regulate kDNA replication in the cell remained poor. The following paragraphs outline several aspects of the control of kDNA replication, including the activation of the replication origins, regulation of the origin binding protein action, and the control of expression of kDNA replication proteins.

Kinetoplast DNA replicates in the trypanosomatid cell during a discrete S phase. The kinetoplast S phase (S_k) and kDNA segregation precedes the nuclear S phase (S_n) and mitosis.[13,67] Kinetoplast segregation is presumably well-coordinated with mitosis.[14] However, studies on the mechanism of cell cycle control[111-117] indicated an apparent uncoupling between the kinetoplast and the nuclear cycle. As the kDNA network replicates at a discrete S-phase, the thousands of origins present in its minicircles and maxicircles have to be regulated by a mechanism, which allows the activation of these origins only during the S phase of the cell cycle. The licensing of chromosomal origins in eukaryotes involves the action of MCM proteins (recently reviewed in refs. 118,119). MCM proteins encoding genes are also present in the trypanosomatids genomes, yet the mechanism that functions in the licensing of kDNA replication origins is unknown.

Regulation of chromosome replication in cells and their subcellular organelles is mediated by interactions of origins of replication with their counterparts, the trans-acting proteins.[120] In kDNA minicircles, two short sequences were associated with the process of replication initiation, a dodecameric sequence GGGGTTGGTGTA (CSB-3, UMS) and a hexameric sequence ACGCCC (within the (CSB-1). Although the mode of activation of kDNA replication origins is unknown, recent studies on the regulation of UMSBP activity through redox signaling (see below), may shed light on the process of activation of minicircle replication initiation.

UMSBP binds specifically the two sequences conserved at the minicircle replication origins, the UMS dodecamer and a 14-mer sequence that contains the core hexamer. Based on its binding to the conserved origin sequences, UMSBP was suggested to play a role during minicircle replication initiation at the replication origin. However, its precise function has yet to be studied. The protein has been purified from *C. fasciculata* and its encoding gene and genomic locus were cloned and analyzed.[79-81,83,84,121] The 116 amino acids protein contains five tandemly arranged CCHC-type zinc-finger motifs. Immunofluorescence analyses demonstrated the dynamic nature of UMSBP localization within the kinetoplast, reaching its maximal level during S-phase, in correlation with the progress in kDNA replication.[76] Structure-function analyses[82] revealed that UMSBP oligomerizes in solution, but binds the origin sequence only in its monomeric form. Furthermore, these analyses indicated that zinc fingers that are involved in the binding of DNA differ from those mediating protein-protein interactions that lead to UMSBP dimerization. Both UMSBP binding to DNA and its dimerization are sensitive to redox potential. Oxidation of UMSBP results in the protein dimerization, mediated by its N-terminal domain, with a concomitant inhibition of its DNA binding activity. UMSBP reduction yields monomers that are active in the binding of DNA, through the protein C-terminal region. *C. fasciculata* tryparedoxins (*Cf*TXNI and II[122-125]) were shown to activate in vitro the binding of an oxidized UMSBP substrate to the DNA (Fig. 3). These results may imply that a cellular redox signaling mechanism may control the binding of UMSBP to the minicircle replication origin. Based on these observations, one may speculate that redox signaling may be involved in the triggering of replication initiation at the minicircle replication origin.

Cycling of the level of expression of kDNA replication proteins with the progress in cell cycle, may serve an important regulatory function during kDNA replication. Dan Ray and his colleagues have observed that the mRNA levels of several nuclear and kDNA replication genes in *C. fasciculata* cycle, as cells progress through the cell cycle,[126-131] reaching their maximal levels at the beginning of S phase, and then decline sharply as DNA synthesis is completed. They showed that cycling is controlled prior to mRNA maturation,[130] and that an octamer sequence [(C/A)AUAGAA(G/A)], located at either the 5' or the 3' untranslated regions (UTRs) of the mRNAs, was involved in the regulation mechanism. Two protein complexes, designated CSBP I and CSBP II, bind specifically to this sequence.[131] This phenomenon has been observed also in *Leishmania infantum*[132] and *Leishmania major*[133] and is presumably shared by other trypanosomatid species.

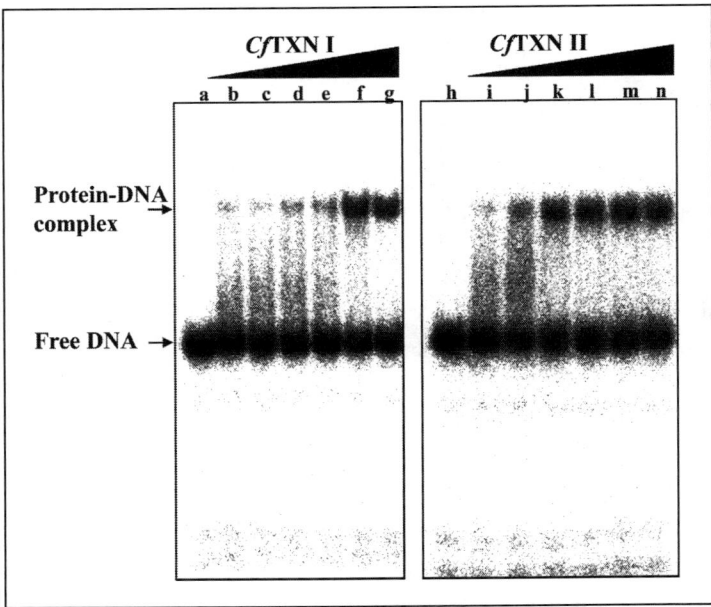

Figure 3. *C. fasciculata* tryparedoxins activate in vitro the binding of UMSBP to UMS DNA. A reactions, which leads from NADPH, through the action trypanothione reductase (TR), to the reduction of trypanothione disulfide (TS_2) to trypanothione ($T[SH]_2$) and subsequently to the reduction of tryparedoxin (TXN), was coupled in vitro to the binding reaction of UMSBP to DNA. The binding of preoxidized UMSBP to an oligonucleotide, representing the origin-associated universal minicircle sequence (UMS), was monitored. Generation of nucleoprotein complexes in the coupled reactions is demonstrated in the electrophoretic mobility shift analyses (EMSA). The two *C. fasciculata* tryparedoxins (*Cf*TXN I and *Cf*TXN II[122-125]) were used. The tryparedoxin reaction included either increasing concentrations of *Cf*TXN I (lanes c-g: 0.01, 0.02, 0.04, 0.06, and 0.1 µM) or *Cf*TXN II (lanes j-n: 0.1, 0.25, 0.5, 0.75, 1 µM). Binding reactions, using 0.6 ng UMSBP that was preoxidized by diamide, were conducted in the presence of 12.5 fmol ^{32}P-labelled UMS DNA. In lanes a and h: no UMSBP and tryparedoxin added; lanes b and i: no tryparedoxin added. Reproduced from: Onn I et al. Redox potential regulates binding of universal minicircle sequence binding protein at the kinetoplast DNA replication origin. Eukaryot Cell 2004; 3(2):277-87;[82] ©2004 with permission of American Society of Microbiology.

Concluding Remarks

The significant advances achieved in recent years in the understanding of the enzymatic system that catalyzes the assembly of the kDNA network in trypanosomatids, was greatly enhanced by using the powerful combination of classical enzymology, coupled with molecular genetics, genomic and proteomic strategies. Nevertheless, several of the basic questions, concerning the structure, function, replication and segregation of kDNA remained unsolved. An intriguing problem is the functional advantage that led to the evolution of this unusual topological catenane. Other interesting mechanisms are the systems that control the accurate scission of the replicated network, the regulation of minicircles' segregation, the functional role of the clusters of replication proteins surrounding the kDNA disk, and the mechanism that 'licenses' kDNA origins to fire and initiate a new round of replication during S phase. Many other questions regarding the details of the replication scheme remained unanswered. Among these are the identity of the topoisomerase II that functions in the prereplication release of minicircles into the KFZ, the mechanism that directs the migration of newly replicated minicircles from the KFZ to the two antipodal attachment sites, the functional rationale for the delayed repair of replicated minicircles, and the mechanism of replication of catenated maxicircles. These and other basic questions on the replication of kDNA will continue to intrigue investigators in this field in the coming years. However, the knowledge gained on the replication mechanisms, proteins, and intermediates, as well as the recent data suggesting that kDNA and mitochondrial functions are not dispensable in the parasites' life cycle, opens new possibilities for the selection of specific targets and the rational designing of drugs against pathogenic trypanosomatids.

Acknowledgements

We are grateful to Dr. Paul T. Englund for permission to use a figure from his published paper and for providing us with the original drawing, and to Dr. Yosef Schlein for helpful comments on the manuscript. Studies in our laboratory were supported, in parts, by grant No. 623 from the Israel Science Foundation (ISF), founded by the Israel Academy of Sciences and Humanities, and grant No. 2001006 from the United State-Israel Binational Science Foundation (BSF), Jerusalem, Israel.

References

1. Klingbeil MM, Drew ME, Liu Y et al. Unlocking the secrets of trypanosome kinetoplast DNA network replication. Protist 2001; 152(4):255-62.
2. Morris JC, Drew ME, Klingbeil MM et al. Replication of kinetoplast DNA: An update for the new millennium. Int J Parasitol 2001; 31(5-6):453-8.
3. Shlomai J. The structure and replication of kinetoplast DNA. Curr Mol Med 2004; 4(6):623-47.
4. Liu B, Liu Y, Motyka SA et al. Fellowship of the rings: The replication of kinetoplast DNA. Trends Parasitol 2005; 21(8):363-9.
5. Lukes J, Guilbride DL, Votypka J et al. Kinetoplast DNA network: Evolution of an improbable structure. Eukaryot Cell 2002; 1(4):495-502.
6. Lukes J, Hashimi H, Zikova A. Unexplained complexity of the mitochondrial genome and transcriptome in kinetoplastid flagellates. Curr Genet 2005; 48(5):277-99.
7. Simpson L, Aphasizhev R, Gao G et al. Mitochondrial proteins and complexes in Leishmania and Trypanosoma involved in U-insertion/deletion RNA editing. RNA 2004; 10(2):159-70.
8. Stuart KD, Schnaufer A, Ernst NL et al. Complex management: RNA editing in trypanosomes. Trends Biochem Sci 2005; 30(2):97-105.
9. Schnaufer A, Domingo GJ, Stuart K. Natural and induced dyskinetoplastic trypanosomatids: How to live without mitochondrial DNA. Int J Parasitol 2002; 32(9):1071-84.
10. Schnaufer A, Panigrahi AK, Panicucci B et al. An RNA ligase essential for RNA editing and survival of the bloodstream form of Trypanosoma brucei. Science 2001; 291(5511):2159-62.
11. Schnaufer A, Clark-Walker GD, Steinberg AG et al. The F1-ATP synthase complex in bloodstream stage trypanosomes is an unusual and essential function. EMBO J 2005; 24(23):4029-40.
12. Das A, Gale Jr M, Carter V et al. The protein phosphatase inhibitor okadaic acid induces defects in cytokinesis and organellar genome segregation in Trypanosoma brucei. J Cell Sci 1994; 107(Pt 12):3477-83.
13. Ploubidou A, Robinson DR, Docherty RC et al. Evidence for novel cell cycle checkpoints in trypanosomes: Kinetoplast segregation and cytokinesis in the absence of mitosis. J Cell Sci 1999; 112(Pt 24):4641-50.
14. Robinson DR, Gull K. Basal body movements as a mechanism for mitochondrial genome segregation in the trypanosome cell cycle. Nature 1991; 352(6337):731-3.
15. Pearson TW, Beecroft RP, Welburn SC et al. The major cell surface glycoprotein procyclin is a receptor for induction of a novel form of cell death in African trypanosomes in vitro. Mol Biochem Parasitol 2000; 111(2):333-49.
16. Welburn SC, Barcinski MA, Williams GT. Programmed cell death in trypanosomatids. Parasitol Today 1997; 13(1):22-6.
17. Perez-Morga D, Englund PT. The structure of replicating kinetoplast DNA networks. J Cell Biol 1993; 123(5):1069-79.
18. Ferguson M, Torri AF, Ward DC et al. In situ hybridization to the Crithidia fasciculata kinetoplast reveals two antipodal sites involved in kinetoplast DNA replication. Cell 1992; 70(4):621-9.
19. Tittawella I. A simple procedure for detecting proteins that bind preferentially to kDNA networks. FEMS Microbiol Lett 1989; 51(3):347-52.

20. Tittawella I. Kinetoplast DNA-aggregating proteins from parasitic protozoan Crithidia fasciculata. FEBS Lett 1990; 260(1):57-61.
21. Tittawella I. Identification of DNA-binding proteins in the parasitic protozoan Crithidia fasciculata and evidence for their association with the mitochondrial genome. Exp Cell Res 1993; 206(1):143-51.
22. Tittawella I, Carlsson L, Thornell LE. Two proteins involved in kinetoplast compaction [published erratum appears in FEBS Lett 1993 Dec 20;336(1):190]. FEBS Lett 1993; 333(1-2):5-9.
23. Xu C, Ray DS. Isolation of proteins associated with kinetoplast DNA networks in vivo. Proc Natl Acad Sci USA 1993; 90(5):1786-9.
24. Xu CW, Hines JC, Engel ML et al. Nucleus-encoded histone H1-like proteins are associated with kinetoplast DNA in the trypanosomatid Crithidia fasciculata. Mol Cell Biol 1996; 16(2):564-76.
25. Lukes J, Hines JC, Evans CJ et al. Disruption of the Crithidia fasciculata KAP1 gene results in structural rearrangement of the kinetoplast disc. Mol Biochem Parasitol 2001; 117(2):179-86.
26. Braly P, Simpson L, Kretzer F. Isolation of kinetoplast-mitochondrial complexes from Leishmania tarentolae. J Protozool 1974; 21(5):782-90.
27. Gull K. The cytoskeleton of trypanosomatid parasites. Annu Rev Microbiol 1999; 53:629-55.
28. Rauch CA, Perez-Morga D, Cozzarelli NR et al. The absence of supercoiling in kinetoplast DNA minicircles. EMBO J 1993; 12:403-11.
29. Chen J, Rauch CA, White JH et al. The topology of the kinetoplast DNA network. Cell 1995; 80(1):61-9.
30. Shapiro TA. Kinetoplast DNA maxicircles: Networks within networks. Proc Natl Acad Sci USA 1993; 90(16):7809-13.
31. Ferguson ML, Torri AF, Perez-Morga D et al. Kinetoplast DNA replication: Mechanistic differences between Trypanosoma brucei and Crithidia fasciculata. J Cell Biol 1994; 126(3):631-9.
32. Simpson L. The mitochondrial genome of kinetoplastid protozoa: Genomic organization, transcription, replication, and evolution. Annu Rev Microbiol 1987; 41:363-82.
33. Stuart K, Feagin JE. Mitochondrial DNA of kinetoplastida. Int Rev Cytol 1992; 141:65-88.
34. Schneider A. Unique aspects of mitochondrial biogenesis in trypanosomatids. Int J Parasitol 2001; 31(13):1403-15.
35. Hancock K, Hajduk SL. The mitochondrial tRNAs of Trypanosoma brucei are nuclear encoded. J Biol Chem 1990; 265(31):19208-15.
36. Schneider A, Marechal-Drouard L. Mitochondrial tRNA import: Are there distinct mechanisms? Trends Cell Biol 2000; 10(12):509-13.
37. Horton TL, Landweber LF. Rewriting the information in DNA: RNA editing in kinetoplastids and myxomycetes. Curr Opin Microbiol 2002; 5(6):620-6.
38. Sturm NR, Simpson L. Kinetoplast DNA minicircles encode guide RNAs for editing of cytochrome oxidase subunit III mRNA. Cell 1990; 61(5):879-84.
39. Fouts DL, Wolstenholme DR. Evidence for a partial RNA transcript of the small circular component of kinetoplast DNA of Crithidia acanthocephali. Nucleic Acids Res 1979; 6(12):3785-804.
40. Rohrer SP, Michelotti EF, Torri AF et al. Transcription of kinetoplast DNA minicircles. Cell 1987; 49(5):625-32.
41. Shlomai J, Zadok A. Kinetoplast DNA minicircles of trypanosomatids encode for a protein product. Nucleic Acids Res 1984; 12(21):8017-28.
42. Singh N, Rastogi AK. Kinetoplast DNA minicircles of Leishmania donovani express a protein product. Biochim Biophys Acta 1999; 1444(2):263-8.
43. Steinert M, Van Assel S. Sequence heterogeneity in kinetoplast DNA: Reassociation kinetics. Plasmid 1980; 3:7-17.
44. Barrios M, Riou G, Galibert F. Complete nucleotide sequence of minicircle kinetoplast DNA from Trypanosoma equiperdum. Proc Natl Acad Sci USA 1981; 78:3323-7.
45. Borst P, Fase-Fowler F, Gibson WC. Kinetoplast DNA of Trypanosoma evansi. Mol Biochem Parasitol 1987; 23:31-8.
46. Ray DS. Conserved sequence blocks in kinetoplast minicircles from diverse species of trypanosomes. Mol Cell Biol 1989; 9(3):1365-7.
47. Chen KK, Donelson JE. Sequences of two kinetoplast DNA minicircles of Tryptanosoma brucei. Proc Natl Acad Sci USA 1980; 77(5):2445-9.
48. Kidane GZ, Hughes D, Simpson L. Sequence heterogeneity and anomalous electrophoretic mobility of kinetoplast minicircle DNA from Leishmania tarentolae. Gene 1984; 27:265-77.
49. Sugisaki H, Ray DS. DNA sequence of Crithidia fasciculata kinetoplast minicircles. Mol Biochem Parasitol 1987; 23(3):253-63.
50. Ponzi M, Birago C, Battaglia PA. Two identical symmetrical regions in the minicircle structure of Trypanosoma lewisi kinetoplast DNA. Mol Biochem Parasitol 1984; 13(1):111-9.
51. Degrave W, Fragoso SP, Britto C et al. Peculiar sequence organization of kinetoplast DNA minicircles from Trypanosoma cruzi. Mol Biochem Parasitol 1988; 27(1):63-70.

52. Vallejo GA, Macedo AM, Chiari E et al. Kinetoplast DNA from Trypanosoma rangeli contains two distinct classes of minicircles with different size and molecular organization. Mol Biochem Parasitol 1994; 67(2):245-53.

53. Kitchin PA, Klein VA, Englund PT. Intermediates in the replication of kinetoplast DNA minicircles. J Biol Chem 1985; 260(6):3844-51.

54. Ntambi JM, Englund PT. A gap at a unique location in newly replicated kinetoplast DNA minicircles from Trypanosoma equiperdum. J Biol Chem 1985; 260(9):5574-9.

55. Birkenmeyer L, Ray DS. Replication of kinetoplast DNA in isolated kinetoplasts from Crithidia fasciculata. Identification of minicircle DNA replication intermediates. J Biol Chem 1986; 261(5):2362-8.

56. Birkenmeyer L, Sugisaki H, Ray DS. Structural characterization of site-specific discontinuities associated with replication origins of minicircle DNA from Crithidia fasciculata. J Biol Chem 1987; 262(5):2384-92.

57. Sheline C, Melendy T, Ray DS. Replication of DNA minicircles in kinetoplasts isolated from Crithidia fasciculata: Structure of nascent minicircles. Mol Cell Biol 1989; 9(1):169-76.

58. Sheline C, Ray DS. Specific discontinuities in Leishmania tarentolae minicircles map within universally conserved sequence blocks. Mol Biochem Parasitol 1989; 37(2):151-7.

59. Ryan KA, Englund PT. Replication of kinetoplast DNA in Trypanosoma equiperdum. Minicircle H strand fragments which map at specific locations. J Biol Chem 1989; 264(2):823-30.

60. Marini JC, Levene SD, Crothers DM et al. Bent helical structure in kinetoplast DNA. Proc Natl Acad Sci USA 1982; 79:7664-8.

61. Kitchin PA, Klein VA, Ryan KA et al. A highly bent fragment of Crithidia fasciculata kinetoplast DNA. J Biol Chem 1986; 261(24):11302-9.

62. Ntambi JM, Marini JC, Bangs JD et al. Presence of a bent helix in fragments of kinetoplast DNA minicircles from several trypanosomatid species. Mol Biochem Parasitol 1984; 12(3):273-86.

63. Clayton DA. Replication of animal mitochondrial DNA. Cell 1982; 28(4):693-705.

64. Clayton DA. Replication and transcription of vertebrate mitochondrial DNA. Annu Rev Cell Biol 1991; 7:453-78.

65. Shadel GS. Yeast as a model for human mtDNA replication. Am J Hum Genet 1999; 65(5):1230-7.

66. Clayton DA. Vertebrate mitochondrial DNA-a circle of surprises. Exp Cell Res 2000; 255(1):4-9.

67. Woodward R, Gull K. Timing of nuclear and kinetoplast DNA replication and early morphological events in the cell cycle of Trypanosoma brucei. J Cell Sci 1990; 95(Pt 1):49-57.

68. Steinert M, Van Assel S. [Coordinated replication of nuclear and mitochondrial desoxyribonucleic acids in "Crithidia luciliae"]. Arch Int Physiol Biochim 1967; 75(2):370-1.

69. Cosgrove WB, Skeen MJ. The cell cycle in Crithidia fasciculata. Temporal relationships between synthesis of deoxyribonucleic acid in the nucleus and in the kinetoplast. J Protozool 1970; 17(2):172-7.

70. Simpson L, Braly P. Synchronization of Leishmania tarantolae by hydroxyurea. J Protozool 1970; 17:511-517.

71. Englund PT. The replication of kinetoplast DNA network in Crithidia fasciculata. Cell 1978; 14:157-168.

72. Englund PT. Free minicircles of kinetoplast DNA in Crithidia fasciculata. J Biol Chem 1979; 254:4895-900.

73. Ryan KA, Shapiro TA, Rauch CA et al. Replication of kinetoplast DNA in trypanosomes. Annu Rev Microbiol 1988; 42:339-58.

74. Melendy T, Sheline C, Ray DS. Localization of a type II DNA topoisomerase to two sites at the periphery of the kinetoplast DNA of Crithidia Fasciculata. Cell 1988; 55:1083-1088.

75. Abu-Elneel K. The initiation of Kinetoplast DNA replication in trypanosomatids: Specific protein-DNA interactions at the replication origin: The Hebrew University of Jerusalem, 2002.

76. Abu-Elneel K, Robinson DR, Drew ME et al. Intramitochondrial localization of universal minicircle sequence-binding protein, a trypanosomatid protein that binds kinetoplast minicircle replication origins. J Cell Biol 2001; 153(4):725-34.

77. Engel ML, Ray DS. The kinetoplast structure-specific endonuclease I is related to the 5' exo/endonuclease domain of bacterial DNA polymerase I and colocalizes with the kinetoplast topoisomerase II and DNA polymerase beta during replication. Proc Natl Acad Sci USA 1999; 96(15):8455-60.

78. Johnson CE, Englund PT. Changes in organization of Crithidia fasciculata kinetoplast DNA replication proteins during the cell cycle. J Cell Biol 1998; 143(4):911-9.

79. Abeliovich H, Tzfati Y, Shlomai J. A trypanosomal CCHC-type zinc finger protein which binds the conserved universal sequence of kinetoplast DNA minicircles: Isolation and analysis of the complete cDNA from Crithidia fasciculata. Mol Cell Biol 1993; 13(12):7766-73.

80. Abu-Elneel K, Kapeller I, Shlomai J. Universal minicircle sequence-binding protein, a sequence-specific DNA-binding protein that recognizes the two replication origins of the kinetoplast DNA minicircle. J Biol Chem 1999; 274(19):13419-26.
81. Avrahami D, Tzfati Y, Shlomai J. A single-stranded DNA binding protein binds the origin of replication of the duplex kinetoplast DNA. Proc Natl Acad Sci USA 1995; 92(23):10511-5.
82. Onn I, Milman-Shtepel N, Shlomai J. Redox potential regulates binding of universal minicircle sequence binding protein at the kinetoplast DNA replication origin. Eukaryot Cell 2004; 3(2):277-87.
83. Tzfati Y, Abeliovich H, Avrahami D et al. Universal minicircle sequence binding protein, a CCHC-type zinc finger protein that binds the universal minicircle sequence of trypanosomatids. Purification and characterization. J Biol Chem 1995; 270(36):21339-45.
84. Tzfati Y, Abeliovich H, Kapeller I et al. A single-stranded DNA-binding protein from Crithidia fasciculata recognizes the nucleotide sequence at the origin of replication of kinetoplast DNA minicircles. Proc Natl Acad Sci USA 1992; 89(15):6891-5.
85. Li C, Englund PT. A mitochondrial DNA primase from the trypanosomatid Crithidia fasciculata. J Biol Chem 1997; 272(33):20787-92.
86. Klingbeil MM, Motyka SA, Englund PT. Multiple mitochondrial DNA polymerases in Trypanosoma brucei. Mol Cell 2002; 10(1):175-86.
87. Melendy T, Ray DS. Novobiocin affinity purification of a mitochondrial type II topoisomerase from the trypanosomatid Crithidia fasciculata. J Biol Chem 1989; 264(3):1870-6.
88. Wang Z, Englund PT. RNA interference of a trypanosome topoisomerase II causes progressive loss of mitochondrial DNA. EMBO J 2001; 20(17):4674-83.
89. Torri AF, Englund PT. Purification of a mitochondrial DNA polymerase from Crithidia fasciculata. J Biol Chem 1992; 267(7):4786-92.
90. Torri AF, Englund PT. A DNA polymerase beta in the mitochondrion of the trypanosomatid Crithidia fasciculata. J Biol Chem 1995; 270(8):3495-7.
91. Torri AF, Kunkel TA, Englund PT. A beta-like DNA polymerase from the mitochondrion of the trypanosomatid Crithidia fasciculata. J Biol Chem 1994; 269(11):8165-71.
92. Hines JC, Engel ML, Zhao H et al. RNA primer removal and gap filling on a model minicircle replication intermediate. Mol Biochem Parasitol 2001; 115(1):63-7.
93. Engel ML, Ray DS. A structure-specific DNA endonuclease is enriched in kinetoplasts purified from Crithidia fasciculata. Nucleic Acids Res 1998; 26(20):4733-4738.
94. Liu Y, Motyka SA, Englund PT. Effects of RNA interference of Trypanosoma brucei structure-specific endonuclease-I on kinetoplast DNA replication. J Biol Chem 2005.
95. Sinha KM, Hines JC, Downey N et al. Mitochondrial DNA ligase in Crithidia fasciculata. Proc Natl Acad Sci USA 2004; 101(13):4361-6.
96. Downey N, Hines JC, Sinha KM et al. Mitochondrial DNA ligases of Trypanosoma brucei. Eukaryot Cell 2005; 4(4):765-74.
97. Saxowsky TT, Choudhary G, Klingbeil MM et al. Trypanosoma brucei has two distinct mitochondrial DNA polymerase beta enzymes. J Biol Chem 2003; 8:8.
98. Chen J, Englund PT, Cozzarelli NR. Changes in network topology during the replication of kinetoplast DNA. EMBO J 1995; 14(24):6339-47.
99. Hoeijmakers JH, Weijers PJ. The segregation of kinetoplast DNA networks in Trypanosoma brucei. Plasmid 1980; 4(1):97-116.
100. Robinson DR, Gull K. The configuration of DNA replication sites within the Trypanosoma brucei kinetoplast. J Cell Biol 1994; 126(3):641-8.
101. Wang Z, Drew ME, Morris JC et al. Asymmetrical division of the kinetoplast DNA network of the trypanosome. EMBO J 2002; 21(18):4998-5005.
102. Ogbadoyi EO, Robinson DR, Gull K. A high-order trans-membrane structural linkage is responsible for mitochondrial genome positioning and segregation by flagellar basal bodies in trypanosomes. Mol Biol Cell 2003; 14(5):1769-79.
103. Soultanas P, Wigley DB. DNA helicases: 'Inching forward' [see comments]. Curr Opin Struct Biol 2000; 10(1):124-8.
104. Robinson DR, Sherwin T, Ploubidou A et al. Microtubule polarity and dynamics in the control of organelle positioning, segregation, and cytokinesis in the trypanosome cell cycle. J Cell Biol 1995; 128(6):1163-72.
105. Drew ME, Englund PT. Intramitochondrial location and dynamics of Crithidia fasciculata kinetoplast minicircle replication intermediates. J Cell Biol 2001; 153(4):735-44.
106. Shapiro TA, Englund PT. The structure and replication of kinetoplast DNA. Annu Rev Microbiol 1995; 49:117-43.
107. Shlomai J, Linial M. A nicking enzyme from trypanosomatids which specifically affects the topological linking of duplex DNA circles. Purification and characterization. J Biol Chem 1986; 261(34):16219-25.

108. Carpenter LR, Englund PT. Kinetoplast maxicircle DNA replication in Crithidia fasciculata and Trypanosoma brucei. Mol Cell Biol 1995; 15(12):6794-803.

109. Hajduk SL, Klein VA, Englund PT. Replication of kinetoplast DNA maxicircles. Cell 1984; 36:483-492.

110. Grams J, Morris JC, Drew ME et al. A trypanosome mitochondrial RNA polymerase is required for transcription and replication. J Biol Chem 2002; 277(19):16952-9.

111. Kumar P, Wang CC. Depletion of anaphase-promoting complex or cyclosome (APC/C) subunit homolog APC1 or CDC27 of Trypanosoma brucei arrests the procyclic form in metaphase but the bloodstream form in anaphase. J Biol Chem 2005; 280(36):31783-91.

112. Tu X, Wang CC. The involvement of two cdc2-related kinases (CRKs) in Trypanosoma brucei cell cycle regulation and the distinctive stage-specific phenotypes caused by CRK3 depletion. J Biol Chem 2004; 279(19):20519-28.

113. Tu X, Wang CC. Pairwise knockdowns of cdc2-related kinases (CRKs) in Trypanosoma brucei identified the CRKs for G1/S and G2/M transitions and demonstrated distinctive cytokinetic regulations between two developmental stages of the organism. Eukaryot Cell 2005; 4(4):755-64.

114. Tu X, Wang CC. Coupling of posterior cytoskeletal morphogenesis to the G1/S transition in the Trypanosoma brucei cell cycle. Mol Biol Cell 2005; 16(1):97-105.

115. Hammarton TC, Engstler M, Mottram JC. The Trypanosoma brucei cyclin, CYC2, is required for cell cycle progression through G1 phase and for maintenance of procyclic form cell morphology. J Biol Chem 2004; 279(23):24757-64.

116. Li Z, Wang CC. Functional characterization of the 11 nonATPase subunit proteins in the trypanosome 19 S proteasomal regulatory complex. J Biol Chem 2002; 277(45):42686-93.

117. McKean PG. Coordination of cell cycle and cytokinesis in Trypanosoma brucei. Curr Opin Microbiol 2003; 6(6):600-7.

118. Blow JJ, Dutta A. Preventing rereplication of chromosomal DNA. Nat Rev Mol Cell Biol 2005; 6(6):476-86.

119. Forsburg SL. Eukaryotic MCM proteins: Beyond replication initiation. Microbiol Mol Biol Rev 2004; 68(1):109-31.

120. Kornberg A, Baker TA. DNA Replication. 2nd ed. San Francisco: Freeman, 1991.

121. Tzfati Y, Shlomai J. Genomic organization and expression of the gene encoding the universal minicircle sequence binding protein. Mol Biochem Parasitol 1998; 94(1):137-41.

122. Montemartini M, Kalisz HM, Kiess M et al. Sequence, heterologous expression and functional characterization of a novel tryparedoxin from Crithidia fasciculata. Biol Chem 1998; 379(8-9):1137-42.

123. Montemartini M, Nogoceke E, Singh M et al. Sequence analysis of the tryparedoxin peroxidase gene from Crithidia fasciculata and its functional expression in Escherichia coli. J Biol Chem 1998; 273(9):4864-71.

124. Montemartini M, Steinert P, Singh M et al. Tryparedoxin II from Crithidia fasciculata. Biofactors 2000; 11(1-2):65-6.

125. Nogoceke E, Gommel DU, Kiess M et al. A unique cascade of oxidoreductases catalyses trypanothione-mediated peroxide metabolism in Crithidia fasciculata. Biol Chem 1997; 378(8):827-36.

126. Hines JC, Ray DS. Periodic synthesis of kinetoplast DNA topoisomerase II during the cell cycle. Mol Biochem Parasitol 1997; 88(1-2):249-52.

127. Pasion SG, Brown GW, Brown LM et al. Periodic expression of nuclear and mitochondrial DNA replication genes during the trypanosomatid cell cycle. J Cell Sci 1994; 107(Pt 12):3515-20.

128. Mahmood R, Hines JC, Ray DS. Identification of cis and trans elements involved in the cell cycle regulation of multiple genes in Crithidia fasciculata. Mol Cell Biol 1999; 19(9):6174-82.

129. Mahmood R, Mittra B, Hines JC et al. Characterization of the Crithidia fasciculata mRNA cycling sequence binding proteins. Mol Cell Biol 2001; 21(14):4453-9.

130. Avliyakulov NK, Hines JC, Ray DS. Sequence elements in both the intergenic space and the 3' untranslated region of the Crithidia fasciculata KAP3 gene are required for cell cycle regulation of KAP3 mRNA. Eukaryot Cell 2003; 2(4):671-7.

131. Mittra B, Sinha KM, Hines JC et al. Presence of multiple mRNA cycling sequence element-binding proteins in Crithidia fasciculata. J Biol Chem 2003; 278(29):26564-71.

132. Hanke T, Ramiro MJ, Trigueros S et al. Cloning, functional analysis and post-transcriptional regulation of a type II DNA topoisomerase from Leishmania infantum. A new potential target for anti-parasite drugs. Nucleic Acids Res 2003; 31(16):4917-28.

133. Zick A, Onn I, Bezalel R et al. Assigning functions to genes: Identification of S-phase expressed genes in Leishmania major based on post-transcriptional control elements. Nucleic Acids Res 2005; 33(13):4235-42.

134. Montemartini M, Kalisz HM, Hecht HJ et al. Activation of active-site cysteine residues in the peroxiredoxin-type tryparedoxin peroxidase of Crithidia fasciculata. Eur J Biochem 1999; 264(2):516-24.

CHAPTER 3

Drugs and Transporters in Kinetoplastid Protozoa

Scott M. Landfear*

Abstract

Kinetoplastid protozoa express hundreds of membrane transport proteins that allow them to take up nutrients, establish ion gradients, efflux metabolites, translocate compounds from one intracellular compartment to another, and take up or export drugs. The combination of molecular cloning, genetic approaches, and the completed genome projects for *Trypanosoma brucei*, *Leishmania major*, and *Trypanosoma cruzi* have allowed detailed functional analysis of various transporters and predictions about the likely functions of others. Thus many opportunities exist to define the biological and pharmacological properties of parasite transporters whose genes were often difficult to identify in the pregenomic era. A subset of these transporters that are essential for parasite viability could serve as targets for novel drug therapies by identifying compounds that interfere with their uptake functions. Other permeases provide routes for uptake of selectively cytotoxic compounds and can thus be useful for delivery of drugs. Drug resistance may develop in strains where such drug uptake transporters are nonfunctional or in parasites that over-express other permeases that export a drug. A summary of recent work on *Leishmania* transporters for glucose and for purines is provided as an example of permeases that are being studied in molecular detail.

Introduction

'Transporters', 'carriers', or 'permeases' are polytopic membrane proteins that mediate the translocation of various compounds across biological membranes. Although some molecules, especially those that are small and relatively nonpolar such as oxygen, nitrogen, and carbon dioxide, can pass through lipid bilayers by diffusion without mediation by proteins, most molecules of biological importance are too large and/or hydrophilic to diffuse across membranes at rates sufficient to support the metabolic needs of cells. Consequently all cells express proteins in various membranes that facilitate the passage of specific substances through the membrane barrier. Most cells need to control the transport of a plethora of compounds across their plasma membranes. Hence ion gradients are often created or modulated, organic and inorganic nutrients must be taken up, products of metabolism sometimes must be exported, and potentially toxic compounds that may have entered the cell can be excreted by specific permeases. Given the large number of tasks that must be performed by transporters, it is perhaps not surprising that the number of proteins dedicated to transport by an organism is relatively large. Thus approximately 5% of the proteins encoded by the genome of the yeast *Saccharomyces cerevisiae*

*Scott M. Landfear—Department of Molecular Microbiology and Immunology, Oregon Health & Science University, Portland, Oregon 97239, USA. Email: landfear@ohsu.edu

Drug Targets in Kinetoplastid Parasites, edited by Hemanta K. Majumder.
©2008 Landes Bioscience and Springer Science+Business Media.

(www.yeastgenome.org) and about 2-2.5% of the proteins encoded by parasitic protozoa such as *Leishmania* and *Trypanosoma* species[1] have been annotated as transporters.

Proteins that mediate transport are often classified into several classes depending upon their mode of action. Consequently, 'channels' are specific pores that allow the flux of large numbers of permeants during the open mode. Channels for ions such as potassium, sodium, calcium, and chloride have been studied extensively,[2] especially in higher eukaryotes. In addition, channels can mediate the flux of uncharged substrates, such as the aquaporins and aquaglyceroporins[3] that allow water and glycerol to pass through membranes at rates much higher than that of nonmediated diffusion. Another class of transport proteins includes those referred to as 'facilitative' or 'equilibrative transporters'. These permeases are thought to function in an 'alternating access' mode[4] in which they shuttle between conformations that present permeant binding sites alternately to the opposite sides of a membrane; their classification as facilitative or equilibrative indicates that they do not concentrate their substrates but simply mediate flux down a thermodynamic gradient. Members of a separate class of carriers are designated 'active transporters'. These proteins couple a source of energy to transport to allow the accumulation of a substrate against a concentration gradient and often play critical roles in allowing cells to attain higher concentrations of molecules inside than outside the plasma membrane, or vice versa. Active transporters may be classified as primary, utilizing chemical energy directly, e.g., by enzymatic cleavage of a high energy compound such as ATP, or secondary, those that couple the thermodynamically favorable downhill flux of one substrate such as sodium ions or protons to the concentrative flux of another substrate such as glucose. Secondary active transporters therefore use energy that has been stored in a concentration gradient that was typically established by a primary active transporter.

Research on a large number of transporters and the completion of increasing numbers of genome projects has led to the realization that many permeases from distinct organisms are related and can be grouped into approximately 550 distinct families based upon sequence similarity and related membrane topology profiles. Thus facilitative glucose transporters from humans are related in sequence and almost certainly in structure to sugar transporters from bacteria and many other organisms. The development of the Transport Classification Database (www.tcdb.org)[5] has provided a valuable resource for cataloging these transporter families and for classifying and analyzing newly discovered permeases.

The explosion of information of membrane transport proteins in general has been accompanied by our increasing knowledge of transporters and their function among the kinetoplastid protozoa. The objective of this chapter is not to provide an exhaustive review of transporters in these parasites but to provide a general overview supplemented by several more detailed examples from research in the laboratories of the author and his collaborators. Given the focus of this monograph, particular emphasis will be given to the interaction of parasite carriers with drugs and to permeases as potential drug targets.

Roles of Membrane Transport Proteins among the Kinetoplastida

Drug Uptake and Resistance
Parasite permeases have attracted considerable interest with regard to their roles in drug delivery and resistance. Transporters can contribute to drug resistance in two fundamentally different ways: by serving as specific conduits for import of drugs and by exporting toxic drugs from the parasite. In the former case, decreased expression or function of the transporter generates the drug resistant state. In the latter case, it is increased expression or activity of the relevant permease that leads to increased resistance to the drug.

Transporters Involved in Drug Uptake
One striking example of transporters that mediate drug uptake involves the TbAT1 adenosine/adenine permease of *T. brucei*. Studies on intact parasites identified a transport activity

designated P2 that mediated the uptake of adenosine, adenine, and arsenical drugs such as melarsoprol.[6] Subsequent studies by the same group revealed that diamidine drugs such as pentamidine were also transported by the P2 permease.[7] Although initially it was not clear why a nucleoside/nucleobase transporter should utilize these disparate drugs as substrates, a common structural motif was identified in 6-amino purines, melaminophenyl arsenicals, and pentamidine, and it has been proposed that this motif is responsible for interaction with the P2 transporter.[8] Subsequently, the *TbAT1* gene was cloned by functional expression of adenosine transport activity in yeast and was shown to confer sensitivity to melaminophenyl arsenicals and encode the P2 transporter.[9] A *TbAT1*-null mutant was generated by targeted gene replacement and shown to be deficient in adenosine-sensitive uptake of pentamidine and melaminophenyl arsenicals.[10] However this null mutant was only marginally resistant to these two drugs, and the authors provided kinetic evidence that two other pentamidine transporters exist, HAPT1 and LAPT1, and that HAPT1 may also mediate the residual uptake of the arsenicals.

Purine analogs including pyrazolopyrimidines such as allopurinol,[11] a hypoxanthine analog, and formycin B,[12] an inosine analog, are selectively toxic to *Leishmania* parasites, because they can be utilized efficiently as substrates for the parasite but not the host purine salvage enzymes. Allopurinol is taken up by a nucleobase transport activity in *L. major*,[13] and this allopurinol transporter has been identified recently at the molecular level (unpublished work of D. Ortiz and S. Landfear) as the LmaNT3 purine nucleobase transporter.[14] Similarly, formycin B is taken up by the LdNT2 inosine/guanosine transporter of *L. donovani*,[15] and the cytotoxic adenosine analog tubercidin is transported by the LdNT1 adenosine/pyrimidine nucleoside permeases.[16] Overall, studies on the purine transporters of kinetoplastid parasites reveal that permeases with physiological functions in nutrient uptake can mediate the import of drugs that are either analogs of the natural substrates or that bear minimal structural similarity to these nutrients.

Another recent example of a natural uptake system that confers drug sensitivity has emerged from studies on the aquaglyceroporin channels of *L. major*.[17] The LmAQP1 channel not only catalyzes flux of water and glycerol across the plasma membrane but also mediates uptake of Sb(III), thought to be the active agent in vivo of antimonial drugs such as Pentostam and Glucantime, and As(III). Overexpression of this channel in various *Leishmania* species stimulates transport of Sb(III) and As(III) and confers hypersensitivity to these metalloids, while deletion of one allele of the *LmAQP1* gene confers increased resistance to these substances. Similar channels have been characterized in *T. brucei*,[18] *T. cruzi*,[19] and *Plasmodium falciparum*,[20] where they may also show promise as targets for chemotherapy.[21]

Transporters Involved in Drug Efflux

Transporters that efflux drugs are also able to confer resistance to cytotoxic compounds. Notable among these permeases are the ATP-Binding-Cassette (ABC) transporters.[22] These proteins typically consist of two membrane domains, encompassing six putative transmembrane helices each, as well as two cytosolic hydrophilic nucleotide binding domains, one of which separates the two membrane domains while the other constitutes the COOH-terminal domain of the protein. ABC transporters have been well studied in humans, where the P-glycoprotein ABC transporter, encoded by the *MDR1* gene, mediates resistance to a variety of anti-cancer drugs. In addition, other members of the family are designated Multidrug Resistance associated Proteins (MRPs). ABC transporters constitute large families of proteins in many organisms, and the genomes of *T. brucei* and *L. major* are predicted to encode 18 and 36 ABC transporters respectively.[1] A number of groups have studied these carriers in various kinetoplastid species over the past decade or more, and the results have been comprehensively reviewed by Klokouzas et al.[23] The PGPA and PGPE proteins[24] have been studied in several *Leishmania* species and shown to mediate resistance against As(III) and Sb(III), at least in some cases by promoting export of metal thiol conjugates. MDR1 homologues have been identified in several *Leishmania* species and shown to confer resistance to organic compounds such as vinblastine.[25,26] Recently, the MDR1 protein from *L. enriettii* was localized to various intracellular organelles that are components of the secretory apparatus,[27] suggesting that the protein pumps substrates into secretory compartments that allow export from the parasite. In *T. brucei*, overexpression of the ABC transporters TbMRPA and TbMRPE results in increased resistance to melarsoprol and berenil respectively. In general, the role of ABC transporters in drug resistance in nature remains to be determined.

Nutrient Transporters

As for other organisms, a plethora of transporters known or suspected to be involved in uptake of nutrients exists among the kinetoplastid protozoa.[1] Particularly prominent among these are several families of amino acid transporters,[28,29] but to date only a few of these permeases have been characterized at both the molecular and functional levels.[30,31] Glucose and related hexoses are major nutrients for kinetoplastid parasites, and the glucose transporters from several species of trypanosomes and *Leishmania* have been characterized at the molecular level.[32] Purines are essential nutrients for all known kinetoplastid parasites, as these and other parasitic protozoa are not able to synthesize the purine ring de novo and must salvage purine nucleosides or nucleobases from their hosts.[33] A variety of purine transporter genes have been cloned and functionally expressed from *Leishmania* and trypanosomes. All such carriers are members of the Equilibrative Nucleoside Transporter (ENT) family (SLC29 family in the Human Genome Organization Database, http://www.bioparadigms.org/slc/intro.asp, or 2.A.57 family in the Transporter Classification Database) that is widely distributed among eukaryotes, but no parasite transporters have been identified that are members of the Concentrative Nucleoside Transporter (CNT) family, sodium or proton coupled permeases present in mammals and bacteria.[34] A large number of other nutrient transporters exist, most of which have not yet been explored in detail. One strategy with therapeutic potential would be to identify among these transporters those that take up essential nutrients and then attempt to develop inhibitors that block the function of these carriers. One example of such an approach comes from the malaria field, where efforts are already underway to develop inhibitors of the single hexose transporter from *P. falciparum*. One high affinity inhibitor of this permease has been shown to be toxic for *P. berghei* forms within the erythrocyte.[35]

Metabolite Exporters

A variety of metabolites generated by catabolic pathways must be exported from cells to maintain homeostasis. One notable example is afforded by the bloodstream forms of *T. brucei*. This stage of the parasite lives in a high glucose environment and metabolizes glucose via the glycolytic pathway to pyruvate. Since the mitochondrion is largely nonfunctional in the bloodstream form and the Krebs cycle is not active, pyruvate is not further metabolized and would build up to toxic levels if it could not be exported from the cell body. A pyruvate transport activity was identified in bloodstream trypanosomes some years ago,[36] and inhibition of this transporter by compound UK5099 caused intracellular pyruvate to accumulate, resulting in a drop in cytosolic pH and parasite lysis.[37] These results demonstrate the principle that inhibition of an essential export permease can be toxic to parasites and that such transporters could serve as drug targets.

Other Transporters

Many other permeases exist among the kinetoplastida, and some of these have also been studied in reasonable detail. A few examples include ion pumps that concentrate ions across the plasma or organellar membranes,[38,39] sugar-nucleotide transporters that translocate these precursors for glycoconjugate biosynthesis into the Golgi apparatus,[40] and folate and pteridine transporters some of which also take up the drug methotrexate.[41,42] The above represent only a few examples among the hundreds of transporters encoded within the genomes of these parasites, and the array of relatively unstudied permeases remains a promising field of inquiry relevant to both the basic biochemistry and physiology of these protozoa and to drug development.

In the following sections I present a synopsis of research on two families of permeases that my laboratory has studied, those for glucose and purines, and of their potential relevance as drug targets or routes for drug uptake.

Leishmania Glucose Transporters

The uptake of glucose and other hexoses has been studied in several species of kinetoplastid parasite,[32] due in part to the importance of these sugars as nutrient sources. Glucose is a particularly important nutrient for bloodstream trypanosomes that are exposed to a high concentration of this sugar and are dependent upon it for survival. Classical biochemical studies on various *Leishmania* species revealed that glucose is taken up[43] and metabolized[44] most robustly in the promastigote stage of the life cycle that lives in the insect vector and is exposed to high concentrations of glucose and fructose from the cleavage of sucrose present in the sandfly sugar meals.[45] Glucose is not however an essential nutrient for promastigotes, as these life cycle forms can take up and oxidize amino acids such as proline to generate metabolic energy.[46] In contrast, intracellular amastigotes are probably in a relatively sugar poor environment within the macrophage phagolysosome.[47] Indeed, they transport and metabolize glucose at an ~20-fold lower level compared to promastigotes and appear to oxidize fatty acids as a principal energy source.[43,44]

The glucose transporters of *L. mexicana* are encoded by three clustered genes (Fig. 1), *LmGT1*, *LmGT2*, and *LmGT3*, that encode distinct isoforms.[48] These proteins are related in sequence and predicted topology (Fig. 2) to facilitative glucose transporters (SLC2 or 2.A.1.1 family) from other eukaryotes,[49] and the isoforms differ from each other in sequence most markedly, but not exclusively, in the NH_2- and COOH-hydrophilic domains that are located on the cytosolic side of the membrane. Several important distinctions exist with regard to these three isoforms. The mRNAs for *LmGT1* and *LmGT3* are expressed at similar levels in both

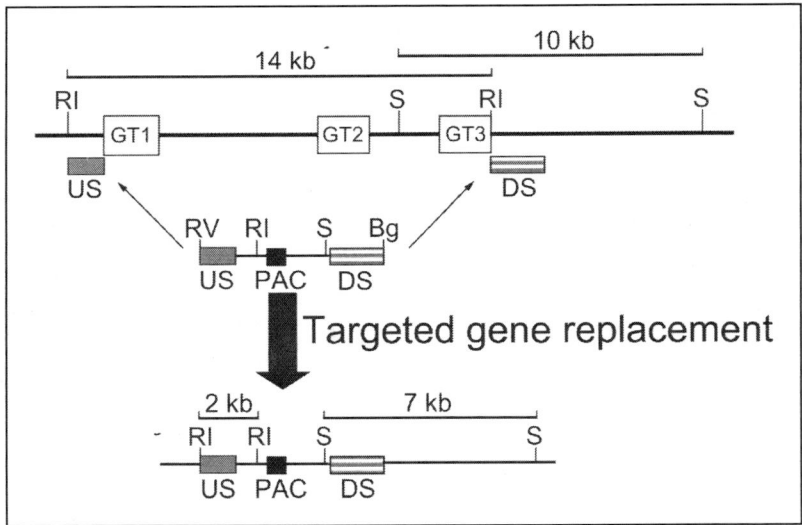

Figure 1. The glucose transporter gene locus in *L. mexicana* encompasses three genes, *GT1*, *GT2* and *GT3*, whose open reading frames are designated by the labeled boxes. The entire gene cluster was deleted by targeted gene replacement using a linear DNA construct containing sequences immediately upstream (US) and downstream (DS) of the cluster and a puromycin acetyl transferase (PAC) selectable marker. A second round of targeted gene replacement employing a nourseothricin resistance marker generated a null mutant at this locus. Sites for the restriction enzymes *Eco*R I (RI), *Sal* I (S), *Eco*R V (RV) and *Bgl* II (Bg) are shown. This figure has been reproduced from reference 51 with permission, ©2003, National Academy of Sciences, U.S.A.

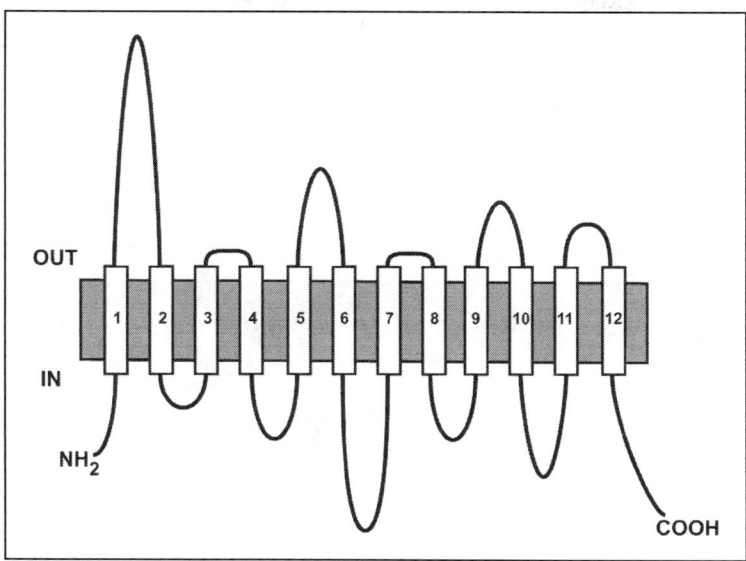

Figure 2. Topology of facilitative glucose transporters in the membrane. The numbered open rectangles represent transmembrane domains, the gray rectangle represents the plasma membrane, and the NH_2 and COOH termini are indicated. Curved black lines represent hydrophilic loops that connect transmembrane domains.

promastigotes and amastigotes, but the mRNA for *LmGT2* accumulates to an ~15-fold higher level in promastigotes compared to amastigotes. In promastigotes, LmGT2 and LmGT3 are localized to the pellicular plasma membrane surrounding the cell body, but remarkably LmGT1 and its homolog from *L. enriettii* ISO1 or LeGT1[50] are preferentially targeted to the flagellar membrane.[51] All three isoforms are able to transport glucose, fructose, mannose, and galactose, although the last is a very low affinity ligand. LmGT2 and LmGT3 are high affinity glucose transporters with K_m values of ~100 μM and ~200 μM respectively, whereas LmGT1 has a higher K_m of ~1 mM. In summary, there are significant differences in function, location, and expression of the three hexose permeases.

To identify biological functions for the *L. mexicana* glucose transporters, the cluster encoding the three genes was deleted by two rounds of homologous gene replacement (Fig. 1) to generate a glucose transporter null mutant designated Δ*lmgt*.[51] This null mutant was viable in the promastigote stage of the life cycle, although it grew to a lower density in glucose containing medium and to populated the sandfly gut with a significantly lower number of parasites compared to the wild type strain. Remarkably and unexpectedly, the glucose transporter null mutants were not viable as amastigotes. Thus the null mutants were unable to establish infections in primary murine macrophages (Fig. 3) or in BALB/c mice (unpublished data of R. Burchmore, G. Coombs, and S. Landfear), and they could not grow as axenic amastigotes in tissue culture medium at pH 5.5 and 32.5°C. These results were initially surprising, as they imply that the glucose transporters are required in the life cycle stage in which they are expressed at the lowest level and in which glucose is used least as an energy source. However, glucose is also utilized as a biosynthetic precursor for making glycoconjugates and complex carbohydrates, and ongoing studies indicate that some of these biosynthetic processes, particularly the accumulation of the storage carbohydrate and virulence factor β-mannan,[52] are reduced in the glucose transporter null mutant (unpublished data of D. Rodriguez-Contreras and S. Landfear). In addition, glucose is metabolized through the pentose phosphate pathway[53] to produce NADPH, that in conjunction with the enzyme trypanothione reductase[54]

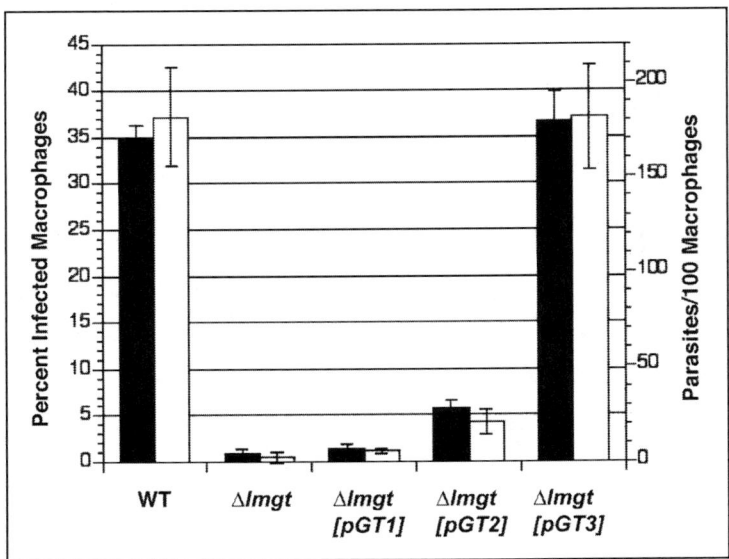

Figure 3. Growth of wild type (WT), glucose transporter null mutant (Δ*lmgt*) and the null mutant complemented with each of the 3 glucose transporters genes on an episome ([pGT1], [pGT2] and [pGT3]). Filled bars represent percent of macrophages infected, and open bars represent parasites per 100 macrophages for the same microscopic fields. Primary peritoneal macrophages were infected with stationary phase promastigotes, and the number of intracellular amastigotes (n = 3, average ± standard deviation) was quantitated 6 days after infection. This figure was reproduced from reference 51 with permission, ©2003, National Academy of Sciences, U.S.A.

maintains the intracellular reducing environment, and ribose phosphate, that is required for the synthesis of nucleotides.[33] We are also studying the effect of deleting the glucose transporter genes on these metabolic pathways.

The observation that glucose transporters are required in the life cycle stage that causes disease suggests that selective inhibitors of these permeases might be of therapeutic value in treating *Leishmania* infections. The fact that the *Leishmania* permeases are quite divergent in sequence from the human homologs[48] suggests that it should be possible to identify inhibitors of the parasite carriers that do not adversely affect the host transporters. This suggestion is also consistent with the fact that inhibitors of the mammalian facilitative glucose transporters, such as phloretin and cytochalasin B, do not efficiently inhibit the *Leishmania* counterparts. As noted above, it has been possible to identify glucose analogs that selectively inhibit the malaria hexose transporters,[35] and in principle a similar strategy may be applied to the *Leishmania* or *T. brucei* hexose permeases.

Leishmania Purine Transporters

The observation that kinetoplastid parasites, unlike their vertebrate hosts, do not synthesize purines de novo but must salvage them using enzymes with distinct substrate specificities suggests that this fundamental distinction between host and parasite might be targeted for therapeutic intervention.[33] The observation that purines and their analogs are imported by a specific set of transporters[55] suggests that this first step of the purine salvage pathway might be one point of attack. Two potential strategies are (i) to employ the purine permeases to take up cytotoxic purine analogs such as allopurinol or formycin B that will selectively kill

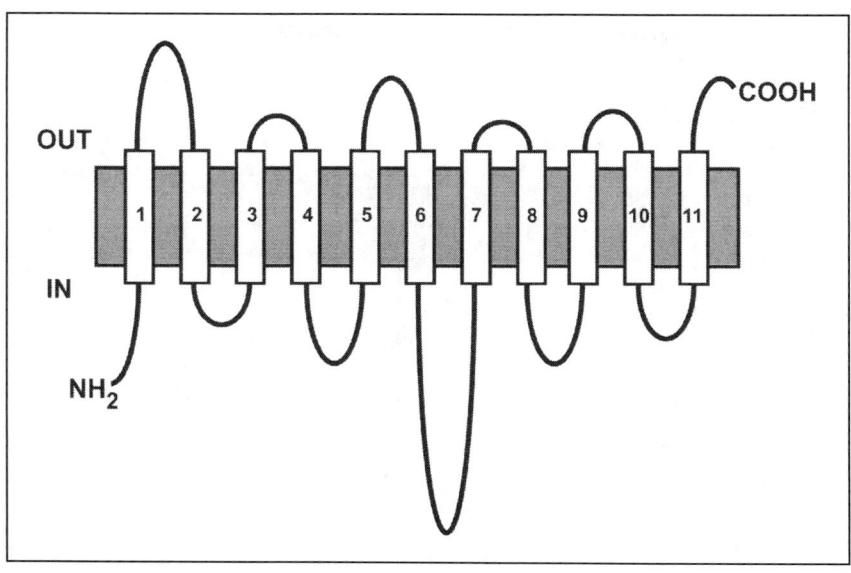

Figure 4. Topology of ENTs in the membrane. The numbered open rectangles represent transmembrane domains, the gray rectangle represents the plasma membrane, and the NH_2 and COOH termini are indicated. This figure has been reproduced with modifications from reference 55 with permission, ©2003, American Society of Microbiology, U.S.A.

the parasites, or (ii) to identify which of the purine permeases are essential and to target them for development of specific transport inhibitors.

Molecular cloning studies in *Leishmania* and trypanosomes have revealed that all known purine transporters in these parasites are members of the ENT family described above. These proteins have 11 predicted transmembrane domains with the NH_2- and COOH termini located on the inside and outside of the plasma membrane respectively (Fig. 4), and this topological model has been tested for the human equilibrative nucleoside transporter hENT1.[56] Four *Leishmania* ENT permeases have been characterized, NT1 (represented by two isoforms, NT1.1 and NT1.2), NT2, NT3 and NT4, with the majority of the work on NT1[16,57-59] and NT2[15,60,61] being performed using the genes from *L. donovani* and the majority of the work for NT3[14] and NT4 (unpublished work of D. Ortiz, M. Sanchez, and S. Landfear) arising from studies on the *L. major* genes. These four transporters are the only members of the ENT family found in the *L. major* genome project (www.genedb.org) and are the only known purine permeases in these parasites. Functional expression has demonstrated that NT1.1 and NT1.2 are adenosine/pyrimidine nucleoside transporters, NT2 is an inosine/guanosine transporter, and NT3 is a hypoxanthine/xanthine/guanine/adenine (purine nucleobase) transporter. Ongoing work on NT4 (unpublished results of D. Ortiz, M. Sanchez, and S. Landfear) suggests that it is a low affinity adenine permease that might have other as yet unidentified ligands.

Studies on these cloned transporter genes have begun to reveal residues and structural domains that are essential for transport function for both NT1 and NT2.[57,59-61] In addition, electrophysiological experiments have confirmed that these parasite permeases, unlike the mammalian ENTs, are secondary active transporters that couple the thermodynamically favorable influx of protons to import of purines.[58] This arrangement presumably allows the parasite to compete effectively for purines with its host and concentrate these essential nutrients in its cytosol.

Drug transport properties of these permeases have been characterized and are those that would be predicted on the basis of their known substrate specificities. Thus NT1.1 and NT1.2 transport the cytotoxic adenosine analog tubercidin,[16] NT2 takes up the inosine analog formycin B,[15] and NT3 mediates the uptake of the hypoxanthine analog allopurinol (unpublished results of D. Ortiz, M. Sanchez, and S. Landfear). Indeed mutants deficient in the *NT1* genes are tubercidin-resistant and those deficient in the *NT2* gene are resistant to formycin B (unpublished results of W. Liu and B. Ullman).

The process of identifying essential *NT* genes by targeted gene deletion is ongoing but currently suggests that the nucleobase but not the nucleoside permeases are required for parasite viability. Thus it has been possible to generate a viable double null mutant of the nucleoside transporter genes, Δ*nt1*/Δ*nt2*, in *L. donovani* (unpublished results of W. Liu and B. Ullman). In addition, null mutants in either nucleobase transporter gene from *L. major*, Δ*nt3* and Δ*nt4*, were viable. However, extensive efforts to generate a double null mutant of the nucleobase transporter genes, Δ*nt3*/Δ*nt4*, resulted only in strains that had undergone the correct targeted integrations but had duplicated a copy of the last gene targeted (unpublished data of D. Ortiz, M. Sanchez, and S. Landfear). This genotype is characteristic of attempts to delete essential genes,[62] as the only parasites that can survive the final gene deletion are those that have acquired an additional copy of the targeted sequence, e.g., by aneuploidy of the targeted chromosome. These observations suggest the possibility that developing selective inhibitors of the NT3 and NT4 permeases might be useful as a combination therapy that could be toxic for the parasites.

Acknowledgements

The author would like to thank present and past members of his laboratory and collaborators from other laboratories for their contributions to the work summarized here. He would also like to acknowledge the National Institutes of Health for support from grants AI25920 and AI44138. I wish to thank Buddy Ullman and Nicola Carter for comments on the manuscript.

References

1. Berriman M, Ghedin E, Hertz-Fowler C et al. The genome of the African trypanosome Trypanosoma brucei. Science 2005; 309(5733):416-422.
2. Hille B. Ion Channels of Excitable Membranes. Sunderland, MA: Sinauer Associates Inc., 2001.
3. King LS, Kozono D, Agre P. From structure to disease: The evolving tale of aquaporin biology. Nat Rev Mol Cell Biol 2004; 5(9):687-698.
4. Kavanaugh MP. Neurotransmitter transport: Models in flux. Proc Natl Acad Sci USA 1998; 95:12737-12738.
5. Busch W, Saier Jr MH. The IUBMB-endorsed transporter classification system. Methods Mol Biol 2003; 227:21-36.
6. Carter NS, Fairlamb AH. Arsenical-resistant trypanosomes lack an unusual adenine/adenosine transporter. Nature 1993; 361:173-175.
7. Carter NS, Berger BJ, Fairlamb AH. Uptake of diamidine drugs by the P2 transporter in melarsen-sensitive and -resistant Trypanosoma brucei brucei. J Biol Chem 1995; 270:28153-28157.
8. Carter NS, Barrett MP, de Koning HP. A drug resistance determinant in Trypanosoma brucei. Trends in Microbiol 1999; 7:469-471.
9. Mäser P, Sütterlin C, Kralli A et al. A nucleoside transporter from Trypanosoma brucei involved in drug resistance. Science 1999; 285:242-244.
10. Matovu E, Stewart ML, Geiser F et al. Mechanisms of arsenical and diamidine uptake and resistance in trypanosoma brucei. Eukaryotic Cell 2003; 2(5):1003-1008.
11. Marr JJ. Purine analogs as chemotherapeutic agents in leishmaniasis and American trypanosomiasis. J Lab Clin Med 1991; 118:111-119.
12. Carson DA, Chang KP. Phosphorylation and anti-leishmanial activity of formycin B. Bioch Biophys Res Comm 1981; 100:1377-1383.
13. Al-Salabi MI, Wallace LJM, de Koning HP. A Leishmania major nucleobase transporter responsible for allopurinol uptake is a functional homolog of the Trypanosoma brucei H2 transporter. Mol Pharmacol 2003; 63:814-820.
14. Sanchez M, Tryon R, Vasudevan G et al. Functional expression and characterisation of a purine nucleobase transporter gene from Leishmania major. Mol Membrane Biol 2004; 21:11-18.

15. Carter NS, Drew ME, Sanchez M et al. Cloning of a novel inosine-guanosine transporter gene from Leishmania donovani by functional rescue of a transport-deficient mutant. J Biol Chem 2000; 275:20935-20941.
16. Vasudevan G, Carter NS, Drew ME et al. Cloning of Leishmania nucleoside transporter genes by rescue of a transport-deficient mutant. Proc Natl Acad Sci USA 1998; 95:9873-9878.
17. Gourbal B, Sonuc N, Bhattacharjee H et al. Drug uptake and modulation of drug resistance in Leishmania by an aquaglyceroporin. J Biol Chem 2004; 279(30):31010-31017.
18. Uzcategui NL, Szallies A, Pavlovic-Djuranovic S et al. Cloning, heterologous expression, and characterization of three aquaglyceroporins from Trypanosoma brucei. J Biol Chem 2004; 279(41):42669-42676.
19. Montalvetti A, Rohloff P, Docampo R. A functional aquaporin colocalizes with the vacuolar proton pyrophosphatase to acidocalcisomes and the contractile vacuole complex of Trypanosoma cruzi. J Biol Chem 2004; 279:38673-38682.
20. Beitz E, Pavlovic-Djuranovic S, Yasui M et al. Molecular dissection of water and glycerol permeability of the aquaglyceroporin from Plasmodium falciparum by mutational analysis. Proc Natl Acad Sci USA 2004; 101(5):1153-1158.
21. Beitz E. Aquaporins from pathogenic protozoan parasites: Structure, function and potential for chemotherapy. Biol Cell 2005; 97(6):373-383.
22. Frelet A, Klein M. Insight in eukaryotic ABC transporter function by mutation analysis. FEBS Lett 2006; 580:1064-1084.
23. Klokouzas A, Shahi S, Hladky SB et al. ABC transporters and drug resistance in parasitic protozoa. Int J Antimicrob Agents 2003; 22(3):301-317.
24. Ouellette M, Legare D, Papadopoulou B. Multidrug resistance and ABC transporters in parasitic protozoa. J Mol Microbiol Biotechnol 2001; 3(2):201-206.
25. Hendrickson N, Sifri CD, Henderson DM et al. Molecular characterization of the ldmdr1 multidrug resistance gene from Leishmania donovani. Mol Biochem Parasitol 1993; 60(1):53-64.
26. Henderson DM, Sifri CD, Rodgers M et al. Multidrug resistance in Leishmania donovani is conferred by amplification of a gene homologous to the mammalian mdr1 gene. Mol Cell Biol 1992; 12(6):2855-2865.
27. Dodge MA, Waller RF, Chow LM et al. Localization and activity of multidrug resistance protein 1 in the secretory pathway of Leishmania parasites. Mol Microbiol 2004; 51(6):1563-1575.
28. Akerman M, Shaked-Mishan P, Mazareb S et al. Novel motifs in amino acid permease genes from Leishmania. Biochem Biophys Res Commun 2004; 325(1):353-366.
29. Bouvier LA, Silber AM, Galvao Lopes C et al. Post genomic analysis of permeases from the amino acid/auxin family in protozoan parasites. Biochem Biophys Res Commun 2004; 321(3):547-556.
30. Hasne MP, Ullman B. Identification and characterization of a polyamine permease from the protozoan parasite Leishmania major. J Biol Chem 2005; 280(15):15188-15194.
31. Shaked-Mishan P, Suter-Grotemeyer M, Yoel-Almagor T et al. A novel high-affinity arginine transporter from the human parasitic protozoan Leishmania donovani. Mol Microbiol 2006; 60(1):30-38.
32. Tetaud E, Barrett MP, Bringaud F et al. Kinetoplastid glucose transporters. Biochem J 1997; 325:569-580.
33. Carter NS, Rager N, Ullman B. Purine and pyrimidine transport and metabolism. In: Marr JJ, Nilsen T, Komuniecki R, eds. Molecular and Medical Parasitology. London: Academic Press, 2003:197-223.
34. Kong W, Engel K, Wang J. Mammalian nucleoside transporters. Curr Drug Metab 2004; 5:63-84.
35. Joet T, Eckstein-Ludwig U, Morin C et al. Validation of the hexose transporter of Plasmodium falciparum as a novel drug target. Proc Natl Acad Sci USA 2003; 100(13):7476-7479.
36. Wiemer EA, Ter Kuile BH, Michels PA et al. Pyruvate transport across the plasma membrane of the bloodstream form of Trypanosoma brucei is mediated by a facilitated diffusion carrier. Biochem Biophys Res Commun 1992; 184(2):1028-1034.
37. Wiemer EA, Michels PA, Opperdoes FR. The inhibition of pyruvate transport across the plasma membrane of the bloodstream form of Trypanosoma brucei and its metabolic implications. Biochem J 1995; 312(Pt 2):479-484.
38. Nolan DP, Revelard P, Pays E. Overexpression and characterization of a gene for a Ca^{+2}-ATPase of the endoplasmic reticulum in Trypanosoma brucei. J Biol Chem 1994; 269:26045-26051.
39. Lu HG, Zhong L, Chang KP et al. Intracellular Ca^{+2} pool content and signaling and expression of a calcium pump are linked to virulence in Leishmania mexicana amazonensis amastigotes. J Biol Chem 1997; 272:9464-9473.
40. Ma D, Russell DG, Beverley SM et al. Golgi GDP-mannose uptake requires Leishmania LPG2. J Biol Chem 1997; 272:3799-3805.

41. Cunningham ML, Beverley SM. Pteridine salvage throughout the Leishmania infectious cycle: Implications for antifolate chemotherapy. Mol Biochem Parasitol 2001; 113:199-213.
42. Richard D, Kundig C, Ouellette M. A new type of high affinity folic acid transporter in the protozoan parasite Leishmania and deletion of its gene in methotrexate-resistant cells. J Biol Chem 2002; 277(33):29460-29467.
43. Burchmore RJS, Hart DT. Glucose transport in promastigotes and amastigotes of Leishmania mexicana: Characterization and comparison with host glucose transporters. Mol Biochem Parasitol 1995; 74:77-86.
44. Hart DT, Coombs GH. Leishmania mexicana: Energy metabolism of amastigotes and promastigotes. Exp Parasitol 1982; 54:397-409.
45. Schlein Y. Sandfly diet and Leishmania. Parasitol Today 1986; 2:175-177.
46. Glew RH, Saha AK, Das S et al. Biochemistry of the Leishmania species. Microbiol Rev 1988; 52:412-432.
47. Burchmore RJ, Barrett MP. Life in vacuoles—nutrient acquisition by Leishmania amastigotes. Int J Parasitol 2001; 31(12):1311-1320.
48. Burchmore RJS, Landfear SM. Differential regulation of multiple glucose transporter genes in the parasitic protozoan Leishmania mexicana. J Biol Chem 1998; 273:29118-29126.
49. Uldry M, Thorens B. The SLC2 family of facilitated hexose and polyol transporters. Pflugers Arch 2004; 447:480-489.
50. Piper RC, Xu X, Russell DG et al. Differential targeting of two glucose transporters from Leishmania enriettii is mediated by an NH_2-terminal domain. J Cell Biol 1995; 128:499-508.
51. Burchmore RJS, Rodriguez-Contreras D, McBride K et al. Genetic characterization of glucose transporter function in Leishmania mexicana. Proc Natl Acad Sci USA 2003; 100(7):3901-3906.
52. Ralton JE, Naderer T, Piraino HL et al. Evidence that intracellular {beta}1-2 mannan Is a virulence factor in Leishmania parasites. J Biol Chem 2003; 278(42):40757-40763.
53. Maugeri DA, Cazzulo JJ, Burchmore RJ et al. Pentose phosphate metabolism in Leishmania mexicana. Mol Biochem Parasitol 2003; 130:117-125.
54. Tovar J, Wilkinson S, Mottram JC et al. Evidence that trypanothione reductase is an essential enzyme in Leishmania by targeted replacement of the tryA locus. Mol Microbiol 1998; 29:653-660.
55. Landfear SM, Ullman B, Carter N et al. Nucleoside and nucleobase transporters in parasitic protozoa. Eukaryotic Cell 2004; 3:245-254.
56. Sundaram M, Yao SYM, Ingram JC et al. Topology of a human equilibrative, nitrobenzylthioinosine (NBMPR)-sensitive nucleoside transporter (hENT1) implicated in the cellular uptake of adenosine and anti-cancer drugs. J Biol Chem 2001; 276:45270-45275.
57. Vasudevan G, Ullman B, Landfear SM. Point mutations in a nucleoside transporter gene from Leishmania donovani confer drug resistance and alter substrate selectivity. Proc Natl Acad Sci USA 2001; 98:6092-6097.
58. Stein A, Vasudevan G, Carter N et al. Equilibrative nucleoside transporter family members from Leishmania donovani are electrogenic proton symporters. J Biol Chem 2003; 278:35127-35134.
59. Valdés R, Vasudevan G, Conklin D et al. Transmembrane domain 5 of the LdNT1.1 nucleoside transporter is an amphipathic helix that forms part of the nucleoside translocation pathway. Biochemistry 2004; 43:6793-6802.
60. Arastu-Kapur S, Ford E, Ullman B et al. Functional analysis of an inosine-guanosine transporter from Leishmania donovani: The role of conserved residues, aspartate 389 and arginine 393. J Biol Chem 2003; 278(35):33327-33333.
61. Arastu-Kapur S, Arendt CS, Purnat T et al. Second-site suppression of a nonfunctional mutation within the Leishmania donovani inosine-guanosine transporter. J Biol Chem 2005; 280(3):2213-2219.
62. Cruz AK, Titus R, Beverley SM. Plasticity in chromosome number and testing of essential genes in Leishmania by targeting. Proc Natl Acad Sci USA 1993; 90:1599-1603.

CHAPTER 4

Selective Lead Compounds against Kinetoplastid Tubulin

R.E. Morgan and K.A. Werbovetz*

Abstract

Kinetoplastid parasites are responsible for the potentially fatal diseases leishmaniasis, African sleeping sickness and Chagas disease. The current treatments for these diseases are far from ideal and new compounds are needed as antiparasitic drug candidates. Tubulin is the accepted target for treatments against cancer and helminths, suggesting that kinetoplastid tubulin is also a suitable target for antiprotozoal compounds. Selective lead compounds against kinetoplastid tubulin have been identified that could represent a starting point for the development of new drug candidates against these parasites.

Introduction

Tubulin is an established target in the chemotherapy of many diseases, as drugs which interact with tubulin are available as treatments for cancer and helminth infections.[1-3] Perhaps the most famous of these tubulin binding drugs is Taxol, which is renowned for both its chemical complexity and remarkable biological effects.[4-7] Tubulin is essential to all eukaryotes, as such there may be many infectious diseases where the development of treatments could be guided by the selection of tubulin as a drug target. Kinetoplastid parasites threaten hundreds of millions of people worldwide, causing leishmaniasis,[8] African sleeping sickness[9] and Chagas disease.[10] Given the inadequate therapeutics available against kinetoplastid parasites, the need for new and effective lead compounds is great.[11] Tubulin is thus a promising target with the potential to provide a new type of therapeutic to aid the treatment of these neglected diseases.[12]

Tubulin

Tubulin is a heterodimeric protein consisting of α and β subunits which polymerises to form microtubules. These microtubules have a number of functions within eukaryotic organisms including chromosome segregation, motility and the maintenance of cellular morphology. The assembly-disassembly process (see Fig. 1) is critical for the proper functioning of microtubules within the cell.[13-15]

The polymerization is environment dependent and requires GTP. Polymer formation is favored at physiological temperatures, while microtubule disassembly occurs at 4°C. The temperature sensitivity of the assembly has been used in order to isolate assembly competent tubulin from a number of species,[16,17] including from *Leishmania*.[18]

*Corresponding Author: K.A. Werbovetz—Division of Medicinal Chemistry and Pharmacognosy, College of Pharmacy, The Ohio State University, 500 West 12th Avenue, Columbus, Ohio 43210, USA. Email: werbovetz.1@osu.edu

Drug Targets in Kinetoplastid Parasites, edited by Hemanta K. Majumder.
©2008 Landes Bioscience and Springer Science+Business Media.

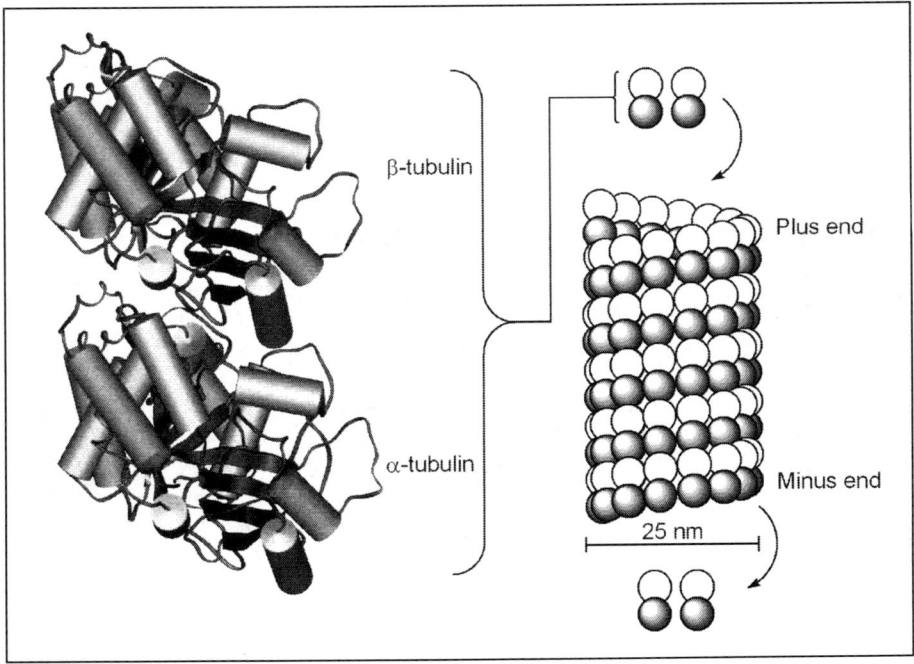

Figure 1. Ribbon diagram of tubulin and a schematic representation of the microtubule. Figure reproduced with permission from: Werbovetz KA. Tubulin as an antiprotozoal drug target. Mini Rev Med Chem 2002; 2:519-529.[12] ©2005 Bentham Science Publishers Ltd.

Tubulin from Protozoan Parasites

In protozoan parasites, tubulin has unique roles.[19] In *Toxoplasma gondii*, for example, an apical organelle known as the conoid is constructed from tubulin.[20] Within kinetoplastid parasites, tubulin is involved in chromosomal segregation and motility, forming the major component of the flagellum.[21] The shape of the parasite is maintained by a network of microtubules which underlie the plasma membrane (the subpellicular microtubules). Tubulin has been identified as being critical for parasite growth, as the addition of anti-tubulin antibodies to *T. brucei* cultures inhibits growth of the parasite.[22]

Tubulin Binding Sites

Known antimitotic agents (see Fig. 2) bind to a number of characterized sites on the tubulin protein. The most well characterized sites include the taxol binding site, the vinca alkaloid domain and the colchicine site.

The taxol (1) binding site has been identified for mammalian tubulin through the use of photoaffinity labelling[23,24] and was later confirmed by electron crystallography.[25,26] Other compounds have been shown to bind to the same site on mammalian tubulin, including epothilone[27,28] and eleutherobin.[29,30] Taxol (1) has been tested against *L. donovani* and *T. cruzi* and was found to block replication in these species, as well as inducing assembly in tubulin isolated from *Leishmania*.[18,31-35] This indicates that these species also contain a binding site for taxol (1) which is similar to that of mammalian tubulin.

Vinca alkaloids such as vinblastine (2) are also known to bind tubulin. Competitive reagents[36] along with fluorescent and fluorescent photoaffinity analogues of vinblastine (2) were synthesised in an attempt to determine the vinca alkaloid binding site.[37,38] Residues 12, 201, 211, 175-213, and 363-379 of β-tubulin were implicated in the binding of this compound as

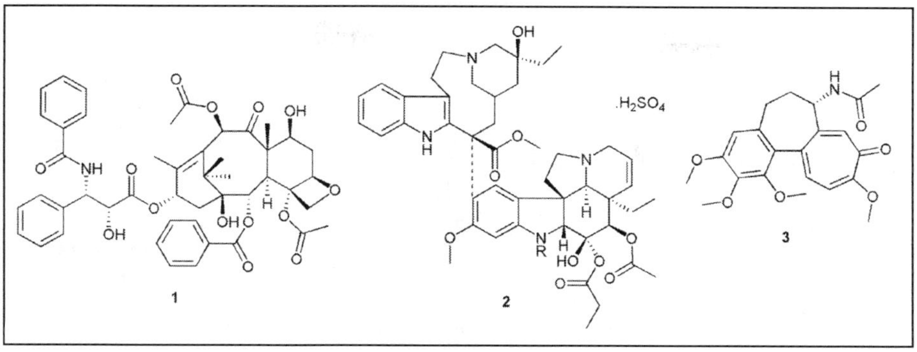

Figure 2. Structures for the tubulin inhibitors taxol (1), vinblastine (2), and colchicine (3).

a result.[36-38] The structure of vinblastine-bound tubulin has since been determined, allowing confirmation of the interaction of vinblastine with some of these residues.[39] Residues 177, 179, 210 and 214 were among those observed to be in contact with vinblastine. Compounds which bind to the vinca domain have shown activity against kinetoplastid parasites, inhibiting the formation of the mitotic spindle and causing parasites to accumulate in the G_2M cell cycle phases.[31,40] The assembly of kinetoplastid tubulin is inhibited by a number of these compounds.[31,41] Identifying new compounds which would bind to the vinca alkaloid domain has the potential to provide promising antikinetoplastid agents. However, given that these compounds are also active against mammalian tubulin, the challenge will be to identify selective compounds. In addition, the chemical complexity of vinca domain agents could also preclude their development as drugs against kinetoplastid parasites.

Colchicine (3) is a well known tubulin binding compound. This compound and its analogues have been used to determine the colchicine binding site through the use of direct, affinity and photoaffinity labelling.[42-45] Peptide fragments consisting of β-tubulin residues 1-46 and 214-241 were directly labeled when colchicine (3) was UV irradiated in the presence of tubulin,[43] and cysteine residues 239 and 354 of β-tubulin were modified by colchicine affinity analogues containing chloroacetyl groups.[44,45] The recent structural determination of colchicine-bound tubulin has shown that the colchicine binding site is buried in the intermediate domain of β-tubulin.[46] This binding site allows colchicine to bind to loop T5 of an adjacent α-tubulin, rationalizing the α-tubulin binding observed in early experiments[42] as well as explaining the observed stabilization of the α/β heterodimer by colchicine.[47] The identified binding site also correlated with the previously reported importance of the residues surrounding 316.[48] Various compounds have been shown to compete with colchicine (3); these compounds presumably bind to β-tubulin in the same region.[12] Colchicine (3) and other compounds known to bind at the cochicine site on mammalian tubulin have little or no effect on kinetoplastid parasites.[18,49] The importance of residue 316 for colchicine activity against mammalian tubulin has been established. *Leishmania* tubulin possesses the sequence ASAL rather than the mammalian VAAV/I in residues 313-316, offering a possible explanation for the observed differences in activity of colchicine. Molecular modeling studies are required to further explore this possibility.

Isolation of Tubulin from Leishmania

The ability to obtain pure, assembly-competent tubulin from kinetoplastid parasites is essential if this protein is to be pursued as a drug target against these organisms. Although expression systems for tubulin have been achieved for a small number of protozoan parasites,[50] to our knowledge, a system for the expression of pure, assembly-competent, heterodimeric tubulin from kinetoplastid parasites has not been reported. However the isolation of such tubulin has been reported from *T. brucei*, *L. amazonensis*, and *Crithidia fasciculata*.[18,49,51] The susceptibility of leishmanial tubulin to a number of mammalian antitubulin agents has been determined.[18]

   L. amazonensis is a causative agent of cutaneous leishmaniasis in humans and as such is considered a BSL-2 pathogen. The isolation of reasonable quantities of assembly-competent leishmanial tubulin requires large numbers of parasites, typically in the region $\sim 5 \times 10^{11}$ cells.[18] Large scale culturing of a pathogenic organism is undesirable. Recent work from our group has demonstrated the successful purification of tubulin from *L. tarentolae*.[52] This *Leishmania* species does not infect humans, its natural host being the Gecko lizard,[53] and as such large scale culture of this organism is more feasible. Additional benefits of isolating tubulin from *L. tarentolae* include the ability to obtain higher cell densities using a less expensive growth medium (Brain Heart Infusion medium versus Schneider's *Drosophila* medium). Clearly *L. tarentolae* is a more desirable organism from which to isolate tubulin, however assays using this protein must be comparable to other *Leishmania* species for this application to be of significance. The amino acid identity observed for tubulin among *Leishmania* species is over 96%, and compounds which are known to inhibit assembly in *L. amazonensis* were found to have comparable activity against *L. tarentolae*.[52] *L. tarentolae* tubulin can thus be used in assays to screen compounds for activity against leishmanial tubulin. This enables drug discovery experiments which would require large amounts of leishmanial tubulin to be pursued with renewed interest.

Oryzalin as a Lead Compound against Kinetoplastid Tubulin

   Trifluralin (4) (see Fig. 3) was known to inhibit microtubule formation in plants and to be selective for plants over mammals.[54] The antileishmainial properties of trifluralin (4) were first identified by Chan et al,[54] who examined the effect of this compound on partially purified leishmanial tubulin and discovered that radiolabled trifluralin (4) showed specific binding to leishmanial tubulin when compared with rat tubulin. Since that initial report, further work has been carried out to examine the effect of trifluralin (4) on homogeneous *Leishmania* tubulin (see below).[18]

   Given the similarity between *Leishmania* and trypanosome tubulin,[12] it is not surprising that trifluralin has also been shown to inhibit trypanosome growth.[55] Chan et al tested trifluralin (4) against *T. brucei* and discovered that proliferation of procyclic trypomastigotes was inhibited by 50% at 6-7 µM concentrations of the compound, with the observed parasite morphologies consistent with tubulin inhibition.[56] Perhaps the most encouraging results from Chan et al were the testing of trifluralin against murine cutaneous leishmaniasis.[56] Trifluralin (4) was applied to the mice using a topical formulation twice a day. The topical formulation consisted of 15% (w/w) trifluralin and 85% carrier. *L. mexicana* lesions were closed after 45 days, with untreated lesions showing an average size of 180 mm^2. In the case of *L. major*, after 30 days 80% of lesions were closed with the remaining showing a 60% reduction in size compared to the group receiving the carrier only. However not all of the treated lesions remained closed, so the results, although promising, are as yet far from optimal.

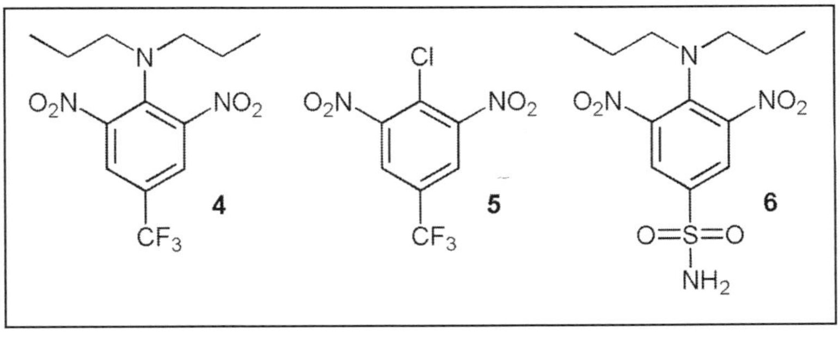

Figure 3. Structures for trifluralin (4), chloralin (5) and oryzalin (6).

A precursor to and occasional contamination in trifluralin (4) is chloralin (5), which has been shown to have improved in vitro antileishmanial properties and also inhibits tubulin polymerisation. This cast understandable concerns over the nature of the activity of trifluralin (4). Investigations into the relative reactivity of these two compounds by Callahan et al showed that chloralin (5) is 100 times more active than trifluralin (4) against *Leishmania*.[57] Experiments to test both trifluralin (4) and chloralin (5) against leishmanial tubulin assembly using purified protein showed chloralin (5) is active against tubulin assembly with an IC_{50} = 22 μM.[18] However, chloralin (5) showed similar activity against rat brain (IC_{50} = 27 μM) and leishmanial tubulin assembly. Trifluralin (4) was found to be inactive against tubulin assembly and less active towards *Leishmania* growth when compared with chloralin. Concerns were raised over the effect of solubility on these results, considering that trifluralin (4) has been shown to have poor solubility properties.[58]

Oryzalin (6) is an analogue of trifluralin (4) where the trifluoromethyl group is replaced with a sulfonamide group. This functional group replacement aids aqueous solubility, making oryzalin (6) a more attractive lead compound. Oryzalin (6) has been shown to inhibit the assembly of 7.5 μM leishmanial tubulin assembly by approximately 54% at a concentration of 25 μM.[59]

SAR Investigations Concerning Oryzalin Analogues and Kinetoplastid Parasites

The synthesis of a number of analogues of oryzalin (6) has been completed, offering a wealth of information as to the structure activity relationship (SAR) surrounding this compound (see Fig. 4 for summary).[59,60] Initial investigations focused mainly on determining the effect of substitutions at position *N*4, with additional compounds providing preliminary information as to the requirement of the sulfonamide and nitro groups at both positions 3 and 5.[59] Reducing the length of the *N*4 substituent either as the ethyl analogue or the mono-alkylated analogue caused a slight decrease in activity. Correspondingly, increasing the length of the *N*4 substitution through the introduction of a butyl, pentyl or hexyl group caused between a 3 to 8 fold increase in potency against *L. donovani* amastigotes. However, substituting with morpholino or pyrrolidino groups caused a decrease in activity, suggesting both that there is a limit to the type of substituent which is tolerated in this position and that restriction of the chain reduces potency.

Replacing the sulfonamide group with either an amide, amidoxime or cyano group causes a loss in activity. Functionalizing the nitrogen of the sulfonamide (*N*1), however, improves the potency of these compounds. The derivative where *N*1 is substituted with a phenyl ring (7) is much more potent than oryzalin (6) against leishmanial tubulin and greater than 10-fold more

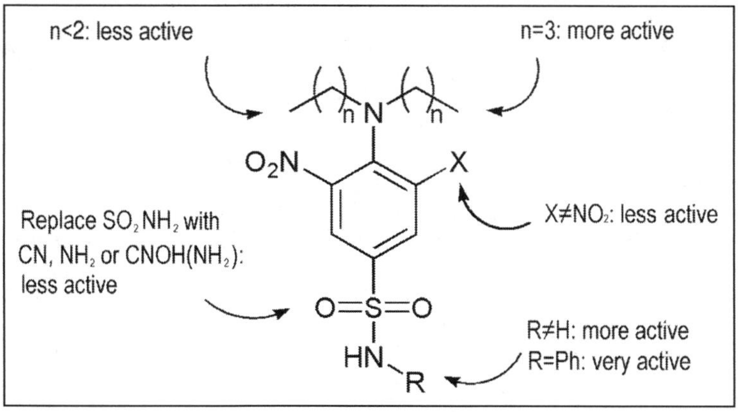

Figure 4. Summary of the SAR data surrounding oryzalin (6).

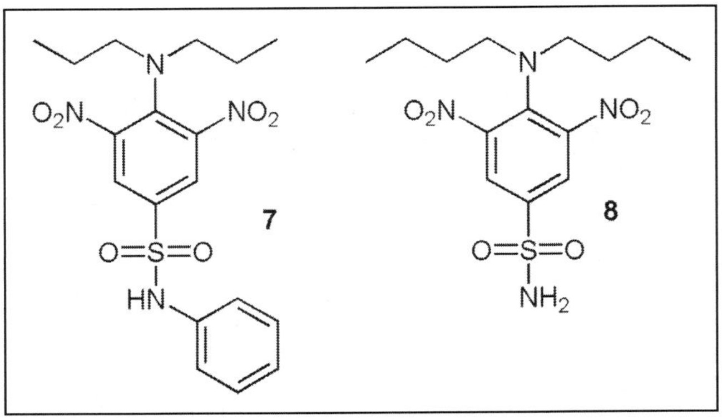

Figure 5. Compounds synthesised and tested for antileishmanial and antitubulin properties. GB-II-5 (7) and GB-II-46 (8).

active against *L. donovani* amastigotes. Synthesis and testing of a derivative with only one nitro group indicated that both nitro groups were required for activity.

Having identified compounds which showed increased potency against leishmanial tubulin and *L. donovani*, compounds 7 and 8 (see Fig. 5), known as GB-II-5 (7) and GB-II-46 (8) were further investigated along with oryzalin for their level of selectivity.[61] Interestingly, GB-II-5 (7) is more active against the African trypanosome than *L. donovani*, with an IC_{50} value of 0.41 μM against *T. b. brucei* variant 221, compared with 5.0 μM against *L. donovani*. GB-II-46 (8) and oryzalin also show greater activity against *T. brucei* compared to *L. donovani*, although these compounds are less potent. GB-II-5 (7) showed the greatest selectivity for kinetoplastid parasites against J774 macrophages and PC3 prostate cell lines, being 71- and 85-fold more active against *T. b. brucei* variant 221 than these two mammalian cell lines, respectively. Complete inhibition of 15 μM leishmanial tubulin assembly was observed with 10 μM and 20 μM GB-II-5 (7) (see Fig. 6A), while the assembly of 15 μM porcine brain tubulin was inhibited by only 17% at a GB-II-5 concentration of 40 μM (see Fig. 6B).

The antiparasitic mechanism of action of these compounds was further investigated by observing their cellular effects on kinetoplastid parasites. An IC_{50} concentration of GB-II-5 (7) was observed to increase the number of *L. donovani* in G_2M and cause the appearance of organisms containing four times the amount of DNA compared to G_1 phase parasites (see Fig. 6C,D). Similar results were observed in *T. brucei* exposed to an IC_{50} concentration of GB-II-5 (7). Treatment with GB-II-5 (7) also caused an increase in the number of *Leishmania* parasites with multiple nuclei and kinetoplasts as assessed by DAPI staining (see Fig. 6E-H). These experiments are entirely consistent with the hypothesis that the antiparasitic mechanism of action of GB-II-5 (7) is tubulin inhibition. However the same mechanistic hypothesis is not supported for oryzalin (6) and GB-II-46 (8), as similar cell cycle effects were not observed with these compounds. Further mechanisms may be contributing to the antiparasitic activity of oryzalin (6) and GB-II-46 (8), which warrant further investigation.

Given the antikinetoplastid potency and selectivity of GB-II-5, further research sought to achieve optimisation of antiparasitic and antimitotic properties by examining the effect of different substituents in position *N*1.[60] A collection of 32 compounds was synthesised, 28 of which varied from oryzalin (6) only by the substitution at the *N*1 atom. Within that collection were a number of analogues containing substituted aromatics. Of these 11 derivatives, none displayed greater potency against *L. donovani* than GB-II-5 (7), although two, a 3'-chloro derivative 9 and a 3',4'-dichloro derivative 10 (see Fig. 7), showed similar activity. However one compound, 3',5'-dichloro derivative 11, showed potency of the same magnitude as GB-II-5 (7) against leishmanial tubulin. Interestingly, of the two compounds with equivalent activity

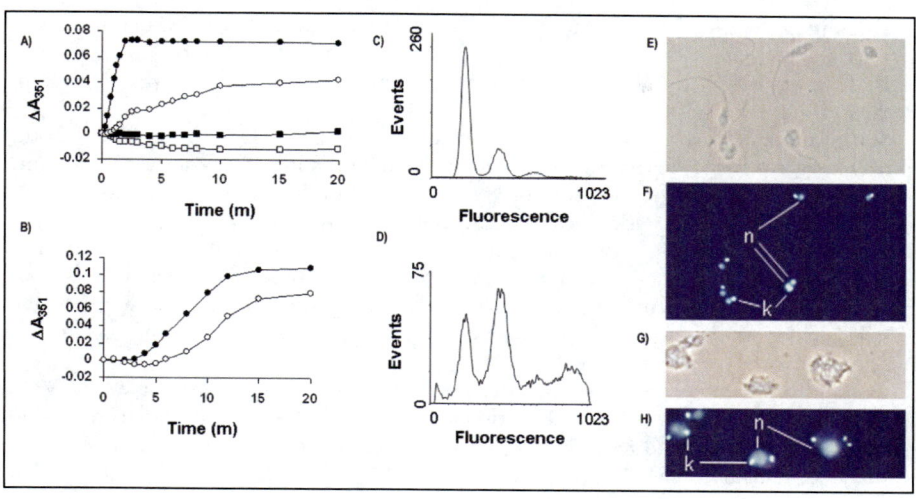

Figure 6. Investigations into the activities of GB-II-5 (7). A) The assembly of 15 μM leishmanial tubulin, assessed at 30°C at 351 nm in the presences of four different concentrations of GB-II-5 (7). **O** = 5, ■ = 10, □ = 20 μM, ● = control. B) Assembly of porcine brain tubulin in the presence (**O**) or absence (●) of 40 μM GB-II-5 (7), assessed at 30°C at 351 nm using 15 μM porcine tubulin. C,D) Cell cycle analysis of *L. donovani* after 48 hour treatments with 1% DMSO, C and GB-II-5 (7), D. E,G) Bright-field microscopy images of *L. donovani* promastigotes after incubation with 1% DMSO, E or GB-II-5 (7), G. F,H) DAPI stained *L. donovani* promastigotes after incubation with 1% DMSO, F or GB-II-5, (7) H, kinetoplast (k), nucleus (n). Figure reproduced with permission from Werbovetz KA et al. Selective antimicrotubule activity of N1-Phenyl-3,5-dinitro-N4,N4-di-n-propylsulfanilamide (GB-II-5) against kinetoplastid parasites. Mol Pharmacol 2003; 64:1325-1333.[61] © 1998-2005, *Molecular Pharmacology Online* by American Society for Pharmacology and Experimental Therapeutics. http://www.molpharm.aspetjournals.org/.

against *L. donovani*, 9 and 10, 9 showed mediocre antitubulin activity but 10 had no observable effect. This suggests that other mechanisms in addition to tubulin inhibition are responsible for the observed activity of compounds 9 and 10.

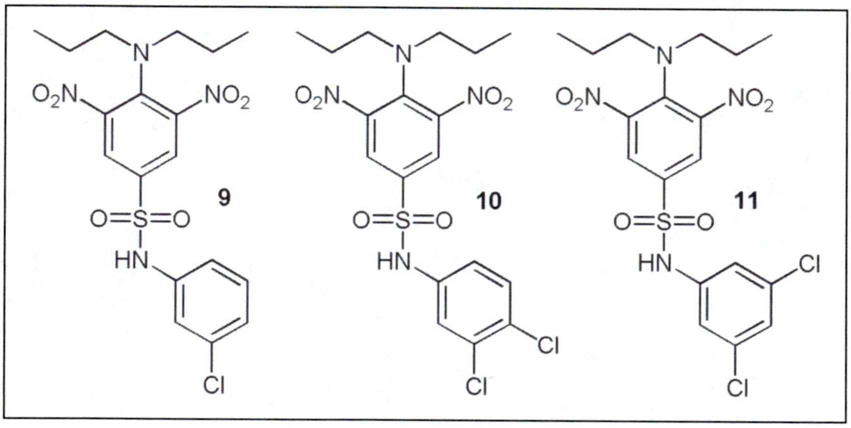

Figure 7. Chlorinated derivatives of GB-II-5 (7).

Examining the effect of the $N1$ aromatic ring substitution pattern revealed that in general derivatives which are *meta* substituted are more potent than their *para* substituted counterparts. This was observed for both the antikinetoplastid and antitubulin activity for chloro, methoxy and methyl substituted derivatives.

Substitution with two phenyl rings at $N1$ caused a substantial reduction in antikinetoplastid activity, suggesting that there are steric restrictions at the tubulin binding site for these compounds. The loss of activity could not solely be attributed to the absence of the sulfonamide proton, as a derivative possessing both a methyl and a phenyl group at $N1$ retained activity, although some reduction in activity was observed. Analogues with an $N1$ alkyl subsituent as opposed to an aromatic ring showed a uniform decrease in activity compared with GB-II-5 (7). However, there was an increase in activity compared to oryzalin (6) for the majority of the $N1$-alkyl analogues. Analogues with hexyl chains lost all activity, which again suggests there is a steric limit at the tubulin binding site.

Combining the two successful substitutions of the phenyl ring at position $N1$ and the dibutyl chains at position $N4$ gave 12, known as GB-II-150 (see Fig. 8). GB-II-150 (12) was the only compound to have potency of the same order of magnitude as GB-II-5 (7) against both *L. donovani* amastigotes, 2.6 μM verses 5.0 μM and leishmanial tubulin 104% inhibition verses 89%.

The activity observed against *T. b. brucei* does not mirror exactly the activity against *L. donovani*. However, GB-II-150 (12) was also the most potent against *T. b. brucei*, possessing an IC_{50} value of 0.12 μM against African trypanosomes compared with 0.41 μM for GB-II-5 (7). GB-II-5 (7), GB-II-150 (12) and 13 showed selectivity towards both *L. donovani* and *T. b. brucei* over J774 macrophages and PC3 prostate cell lines. GB-II-150 (12) displays the most impressive selectivity, with a 100 fold difference in the IC_{50} values for *T. b. brucei* over J774 macrophages and PC3 prostate cell lines.

GB-II-150 (12) is the most successful derivative identified thus far, with both improved potency and selectivity. This compound also causes the G_2M accumulation of *T. b. brucei* when the parasite is exposed to near IC_{50} concentrations of 12, consistent with the hypothesis that the likely target for GB-II-150 (12) in kinetoplastid parasites is tubulin.

GB-II-5 (7) and GB-II-150 (12) were tested against *T. b. brucei* in mice. The compounds were given at a dose of 20 mg/kg/day i.p. for 4 days. No significant increase in survival rate or toxicity was observed. One possible explanation for the poor results in vivo could be poor bioavailability and/or rapid metabolism of the compounds.

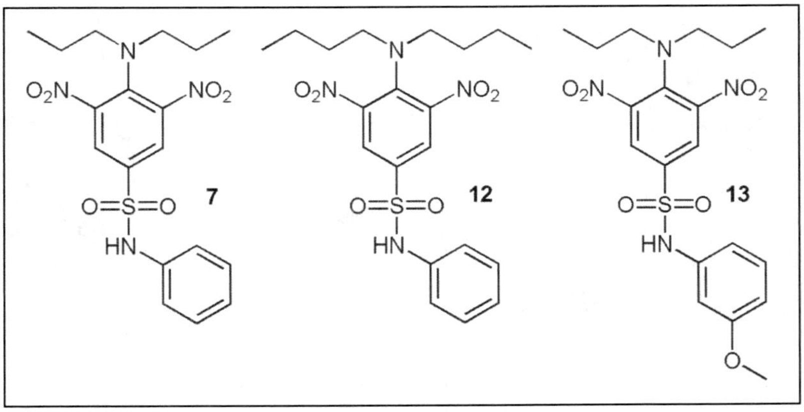

Figure 8. Compounds with the greatest antikinetoplastid selectivity—GB-II-5 (7), GB-II-150 (12) and 13.

Metabolism Studies and New Analogues

Recent studies with GB-II-150 (12) have shown that the compound is extensively metabolised in male Sprague-Dawley rats.[62] The oral bioavailability of the compound was shown to be zero. When administered intravenously, GB-II-150 (12) had a half life of 170 min and a clearance of 31.5 ml/min/kg. Together with the data obtained in the animal model of African trypanosomiasis,[60] these results suggested that metabolism could be a serious issue with these compounds. This hypothesis is further supported by in vitro studies showing that less than 5% of GB-II-150 (12) remained intact after a 60 min incubation with rat liver S9 fractions. The metabolism of GB-II-150 (12) included *N*1-ring oxidation, *N*4-alkane oxidation, *N*4-oxidation and nitro reduction. Eighteen metabolites were identified from in vitro studies and eleven distinct metabolites were observed from in vivo studies. Synthesis of three of the proposed metabolites, M17 (14), M18 (15) and M29 (16) (see Fig. 9), followed by LCMS analysis confirmed their identity and biological assays showed that they were significantly less active than the parent compound. Thus, there are clear indications that metabolism of these types of compounds is an obstacle to their further development as antiparasitic drugs.

Ongoing work focuses on the stabilization of such compounds to metabolic inactivation while maintaining antiparasitic potency and selectivity. Seventeen new dinitroaniline sulfonamide and eleven new benzamide analogs of these leads have recently been reported.[63] Although the benzamides showed little activity, nine of the new sulfonamides displayed in vitro IC_{50} values under 500 nM against African trypanosomes, and the most active antikinetoplastid compounds also inhibited the in vitro assembly of purified leishmanial tubulin with potencies similar to that of GB-II-150 (12). The absence of *N*4 alkyl groups (compound 16) or the presence of bulky dibenzyl substituents at the *N*4 position (compound 17, see Fig. 10) stabilize the analogues to metabolism compared with GB-II-150 (12). Unfortunately, 16 and 17 display low activity against kinetoplastid parasites. Several of the potent compounds, such as 18, are rapidly degraded by rat liver S9 fractions in vitro. However, TG-II-36 (19) displays an IC_{50} value of 260 nM against African trypanosomes in vitro and is more stable than GB-II-150 (12) in the in vitro metabolism assay. The placement of the OH group in the *meta* position on the *N*1 aromatic ring may allow for a critical hydrogen bonding interaction at the tubulin binding site, minimizing any unfavorable steric interactions. Based on in vitro antikinetoplastid activity, selectivity, and metabolic stability, TG-II-36 (19) emerges as a candidate for further evaluation against kinetoplastid parasites. Taken together, these results highlight the importance of the *N*1 and *N*4 substitutions to antimitotic antikinetoplastid activity and the susceptibility of the *N*4 alkyl substituents to metabolism. Increased hydrophilicity of the compounds, as in TG-II-36 (19), may lower their affinity for cytochrome P450 enzymes and lead to improved in vivo properties.

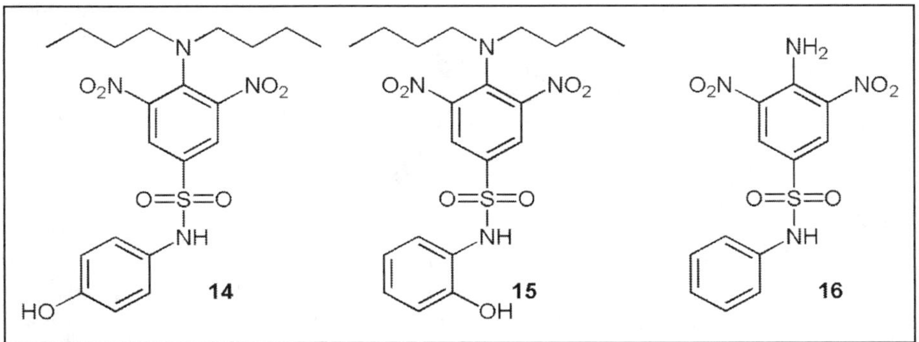

Figure 9. Identified metabolites of GB-II-150. M17 (14) and M18 (15) were identified from the in vitro investigations, while M29 (16) was identified from the in vivo studies.

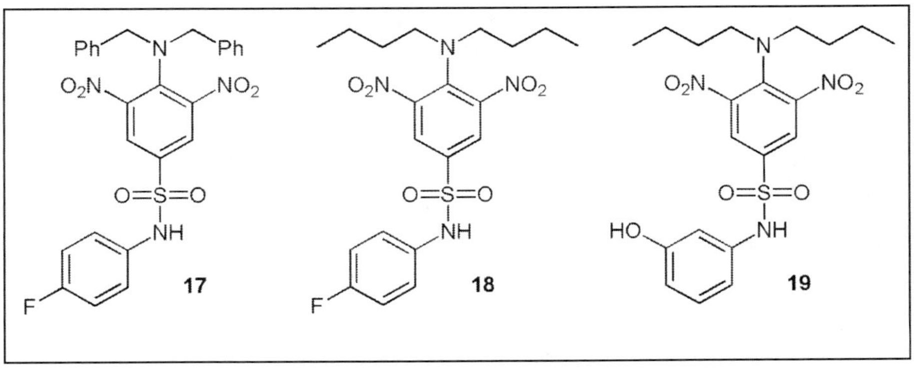

Figure 10. New analogues of GB-II-5 (7) and GB-II-150 (12). TG-II-36 (19) displays both potent antimitotic antikinetoplastid activity and increased in vitro metabolic stability compared to GB-II-150 (12).

Binding Studies with Tubulin from Other Protozoan Parasites

Although the focus of this chapter is on the kinetoplastid parasites, research on *T. gondii*, another protozoan parasite, has been performed which is of interest and could prove to be particularly useful in the progression of these compounds as leads against parasitic tubulin.[64] Morrissette et al used information gained from mutation studies in combination with molecular modeling to identify the likely binding site of oryzalin (6) on *T. gondii* tubulin.

Oryzalin (6) resistant *Toxoplasma* lines were found to have numerous point mutations in the single α-tubulin gene which conferred resistance. This suggests that oryzalin (6) binds to α-tubulin. More specifically, using docking simulations the binding site was identified as being beneath the N loop of α-tubulin and composed of Arg2, Glu3, Val4, Trp21, Phe24, His28, Ile42, Asp47, Arg64, Cys65, Thr239, Arg243 and Ph244. Several of the observed mutations were in this region (at Phe24, His28, Thr239 and Arg243). Other mutations at the core of α-tubulin were also observed and were proposed to affect the dinitroaniline binding site by changing tubulin's conformation. This binding site model provides a possible rationale of the mechanism of oryzalin (6) and its derivatives. Interactions between the N loop and the M loop regions of tubulin allow lateral adhesion. The binding of oryzalin (6) to the area beneath the N loop may disrupt this interaction and so destabilize the microtubules. Disruption in this region has been identified in the mechanism of another compound, Taxol (1). Taxol (1) is known to stabilize microtubules by reinforcing M-N loop interactions in β-tubulin.[26] Docking oryzalin (6) to bovine tubulin revealed a significantly lower binding affinity (>50 fold), which is consistent with the observed selectivity of oryzalin (6) to protozoan parasite tubulin over mammalian tubulin.[64]

Identifying this probable binding site of oryzalin (6) in *T. gondii* tubulin enables rational design approaches to the synthesis of not only analogues of oryzalin (6) but also potentially new lead compounds against *T. gondii*. Having established the model for *T. gondii* tubulin, work is now underway to model leishmanial tubulin with the hopes of facilitating future drug discovery efforts.

Aromatic Thiocyanates as Lead Compounds

The aromatic thiocyanate 2,4-dichlorobenzylthiocyanate (DCBT) (20) (see Fig. 11) is a known inhibitor of mammalian tubulin polymerisation.[65] DCBT (20) has been shown to react at a specific cysteine residue of mammalian β-tubulin, Cys239.[66] Another thiocyanate, WR85915 (21), has been tested against *Leishmania* tubulin assembly. Interestingly, initial studies indicated that although WR85915 (21) inhibits in vitro polymerisation of tubulin in

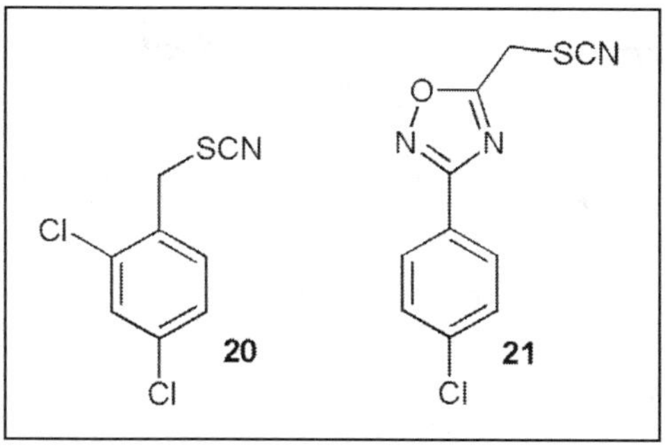

Figure 11. Aromatic thiocyanate tubulin assembly inhibitors. DCBT (20) and WR85915 (21).

Leishmania, this compound does not appear to act through a tubulin mechanism in live parasites.[31] This was concluded from the observation that incubation of WR85915 (21) with *L. donovani* did not cause the usual cell types observed with other tubulin inhibitors and the parasites did not accumulate in G_2M phase.[31] WR85915 (21) must therefore be acting on other cellular target(s) besides or in addition to tubulin.

Antikinetoplastid compounds containing a thiocyanate functional group have been reported elsewhere. Thiocyanate containing analogues of fenoxycarb (22) (see Fig. 12) were identified as inhibitors of *T. cruzi.*[67] 23 was identified as a lead compound with an IC_{50} value of 2.2 μM against *T. cruzi* epimastigotes.[67] Synthesis of analogues around 23 identified the more potent 24, with an IC_{50} of 0.87 μM. However, the activity of these compounds towards tubulin is unknown and one proposed mechanism of action is via the ergosterol biosynthetic pathway.

Given the observed activity of WR85915 (21) and the further indications that thiocyanate containing compounds possess antikinetoplastid properties, a collection of 14 analogues of WR85915 (21) were synthesised.[68] The thiocyanate functional group was essential for antikinetiplastid activity, consistent with the analogous requirement for the activity of DCBT (20) against mammalian tubulin.[69] Testing of WR85915 (21) in vivo against BALB/c mice showed that treatment with 50 mg/kg in five daily oral doses suppressed liver parasitemia by 61%, however a slight weight loss was also observed. Intravenous doses of 5 mg/kg did not cause any toxic effects with approximately 48% suppression observed. These effects were similar to those obtained with a five day subcutaneous treatment of 15 mg/kg Pentostam, a compound currently used in the treatment of leishmaniaisis.

Although DCBT (20) is known to covalently modify mammalian tubulin with selectivity, the mechanism of action of aromatic thiocyanates against kinetoplastid parasites is unclear. WR85915 (21) is an interesting antileishmanial lead compound in its own right given its oral activity versus *L. donovani* in vivo. Such studies indicate that aromatic thiocyanates possess acceptable pharmacokinetic properties. Further work with aromatic thiocyanates may be useful in identifying compounds that selectively interact with parasite tubulin and interfere with the function of this critical protein in the cell.

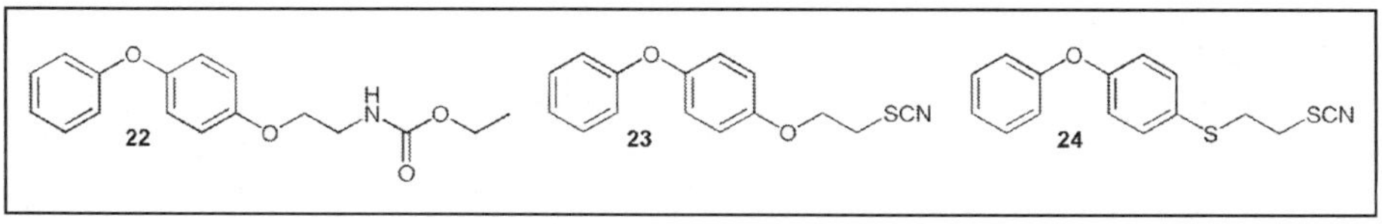

Figure 12. Fenoxycarb (22) and derivatives.

Generating New Leads

While the lead compounds mentioned earlier offer great promise for the development of effective antileishmanial agents, there is still the need to identify new compounds which can function as selective antitubulin agents. High throughput screening is an established technique for the selection of compounds from large libraries on the basis of a desired property or biological activity. Screening a compound library for antitubulin compounds should enable the identification of novel and selective lead molecules.

Screening against Tubulin from Leishmania

High throughput screens (HTS) have not been used routinely in parasitology.[70] With the majority of the research in parasitology based in academic institutions, it is only now as this technique is becoming more accessible to the academic that the number of screens in this area is beginning to increase. Within the last year, two HTS using unbiased compound collections against protozoan parasites have been published.[71,72] One of these screens was carried out against the kinetoplastid parasite *L. tarantole*, as St. George et al screened a 15,000 compound library for in vitro growth inhibition of *L. tarantolae*.[72] The NCI database was used to select out those compounds known to be cytotoxic and those with physical properties not amenable to drug development. Secondary assays were then used to further refine the hit set, including assays for potency, antifungal and antibacterial activities. From these assays 43 compounds were selected which were then tested against *L. major*-infected murine macrophages. Three compounds were identified as potent at clearing infected macrophages.

Screening for compounds which induce mitotic changes has been very successful,[73,74] with the identification of monastrol being one example.[74] The concept of screening against mamalian tubulin has been proposed in the literature, although no actual screen as been published.[75] In addition to discussions around screening against tubulin assembly,[75] another assay which used tubulin-dependent changes in cellular morphology to identify tubulin polymerization inhibitors has been presented.[76] The principle of this assay involved the study of rat cells treated with dibutyryl-cAMP (db-cAMP). db-cAMP causes an alteration in the morphology of C6 cells, prior to addition they exhibit glial (fibroblastic) morphology with elongated stoma with two or more processes. After addition of db-cAMP they become spherical, at which point they can be easily aspirated off culture plates. This alteration in morphology is caused by tubulin polymerization and compounds which inhibit tubulin polymerization stop these morphological changes. Taxol and latrunculin were used as controls in this assay and both did not inhibit db-cAMP induced alteration in the cellular morphology. Treatment of rat glioma cells with vinblastine, vincristine, colchicine and curacin A inhibited the morphological changes, resulting in the cells remaining attached to the 96 well plate surface. This assay could potentially be used in the context of a high throughput screen as proposed by the authors.

Development of a high throughput screen against leishmanial tubulin offers the opportunity to identify new and exciting anti-leishmanial lead compounds. With the availability of tubulin from the nonpathogenic, more readily cultured *L. tarentolae* the large scale production of protein required for a high throughput screen becomes more feasible. Challenges still remain in order to optimize the assay for a large scale high throughput approach, but we feel that such a screen employing this assay has the potential to make a significant impact on antikinetoplastid drug discovery and development.

Conclusion

The demonstrated susceptibility differences between kinetoplastid and mammalian tubulin make this protein not only an attractive but also a feasible target for antikinetoplastid compounds. Oryzalin derivatives GB-II-5 (7), GB-II-150 (12), and TG-II-36 (19) are promising lead compounds showing potent and selective antimitotic effects against kinetoplastid parasites. The development of a model for the dinitroaniline binding site on leishmanial tubulin would permit the application of structure-based drug discovery techniques in order to identify novel compounds that bind selectively to parasite tubulin. With the ability to isolate large quantities of tubulin from the nonpathogenic *L. tarentolae*, investigations into this promising antikinetoplastid target may soon include high throughput screening. Given the critical functions of tubulin in eukaryotic cells, the known differences between mammalian and kinetoplastid proteins in susceptibility to ligands, and the existing opportunities to carry out traditional medicinal chemistry, structure-based drug discovery, and high throughput screening on this target, the exploration of tubulin for the development of new antikinetoplastid drugs appears to be an extremely promising approach.

Acknowledgements

We would like to thank previous and current members of the Werbovetz lab as well as past and present collaborators for their important contributions on this project. Financial support for this work was provided by NIH grant AI061021 (to K.A.W.).

References

1. Pellegrini F, Budman DR. Review: Tubulin function, action of antitubulin drugs, and new drug development. Cancer Invest 2005; 23:264-273.
2. Zhou J, Giannakakou P. Targeting microtubules for cancer chemotherapy. Curr Med Chem - Anti-Cancer Agents 2005; 5:65-71.
3. Martin RJ, Robertson AP, Bjorn H. Target sites of anthelmintics. Parasitol 1997; 114:S111-S124.
4. Nicolaou KC, Yang Z, Liu JJ. Total synthesis of taxol. Nature 1994; 367:630-634.
5. Srivastava V, Negi AS, Kumar JK et al. Plant-based anticancer molecules: A chemical and biological profile of some important leads. Bioorg Med Chem 2005; 13:5892-5908.
6. Wilson L, Miller HP, Farrell KW et al. Taxol stabilization of microtubules in vitro: Dynamics of tubulin addition and loss at opposite microtubule ends. Biochemistry 1985; 24:5254-5262.
7. Arregui L, Muñoz-Fontela C, Serrano S et al. Direct visualization of the microtubular cytoskeleton of ciliated protozoa with a fluorescent taxiod. J Eukaryot Microbiol 2002; 49:312-318.
8. http://www.who.int/leishmaniasis/burden/en/
9. http://www.who.int/tdr/diseases/tryp/diseaseinfo.htm
10. http://www.who.int/ctd/chagas/burdens.htm
11. Croft SL, Barrett MP, Urbina JA. Chemotheraphy of trypanosomiases and leishmaniasis. Trends Parasitol 2005; 21:508-512.
12. Werbovetz KA. Tubulin as an antiprotozoal drug target. Mini Rev Med Chem 2002; 2:519-529.
13. Downing KH, Nogales E. Tubulin structure: Insights into microtubule properties and functions. Curr Opin Struct Biol 1998; 8:785-791.
14. Kline-Smith SL, Walczak CE. Mitotic spindle assembly and chromosome segregation: Refocusing on microtubule dynamics. Mol Cell 2004; 15:317-327.
15. Amos LA, Schlieper D. Microtubules and maps. Adv Protein Chem 2005; 71:257-298.
16. Castoldi M, Popov AV. Purification of brain tubulin through two cycles of polymerization-depolymerization in a high-molarity buffer. Protein Expr Purif 2003; 32:83-88.
17. Fourest-Lieuvin A. Purification of tubulin from limited volumes of cultured cells. Protein Expr Purif 2006; 45:183-190.
18. Werbovetz KA, Brendle JJ, Sackett DL. Purification, characterisation and drug susceptibility of tubulin from Leishmania. Mol Biochem Parasitol 1999; 98:53-65.
19. Gull K. Protist tubulins: New arrivals, evolutionary relationships and insights to cytoskeletal function. Curr Opin Microbiol 2001; 4:427-432.
20. Hu K, Roos DS, Murray JM. A novel polymer of tubulin forms the conoid of Toxoplasma gondii. J Cell Biol 2002; 156:139-1050.
21. Kohl L, Gull K. Molecular architecture of the trypanosome cytoskeleton. Mol Biochem Parasitol 1998; 93:1-9.
22. Lubega GW, Ochola OK, Prichard RK. Trypanosoma brucei: Anti-tubulin antibodies specifically inhibit trypanosome growth in culture. Exp Parasitol 2002; 102:134-142.
23. Rao S, Krauss NE, Heerding JM et al. 3'-(p-Azidobenzamido)taxol photolabels the N-terminal 31 amino acids of β-tubulin. J Biol Chem 1994; 269:3132-3134.
24. Rao S, Orr GA, Chaudhary AG et al. Characterization of the taxol binding site on the microtubule. 2-(m-Azido benzoyl)taxol photolabels a peptide (amino acids 217-231) of tubulin. J Biol Chem 1995; 270:20235-20238.
25. Nogales E, Wolf SG, Downing KH. Structure of the αβ tubulin dimer by electron crystallography. Nature 1998; 391:199-203.

26. Löwe J, Li H, Downing KH et al. Refined structure of αβ-tubulin at 3.5 Å resolution. J Mol Biol 2001; 313:1045-1057.
27. Bollag DM, McQueney PA, Zhu J et al. Epothilones, a new class of microtubule-stabilizing agents with a taxol-like mechanism of action. Cancer Res 1995; 55:2325-2333.
28. Giannakakou P, Gussio R, Nogales E. A common pharmacophore for epothilone and taxanes: Molecular basis for drug resistance conferred by tubulin mutations in human cancer cells. Proc Natl Acad Sci USA 2000; 97:2904-2909.
29. Long BH, Carboni JM, Wasserman AJ et al. Eleutherobin, a novel cytotoxic agent that induces tubulin polymerisation, is similar to paclitaxel (taxol). Cancer Res 1998; 58:1111-1115.
30. Jiménez-Barbero J, Amat-Guerri F, Snyder JP. The solid state, solution and tubulin-bound conformations of agents that promote microtubule stabilization. Curr Med Chem - Anti-Cancer Agents 2002; 2:91-122.
31. Havens CG, Bryant N, Asher L et al. Cellular effects of leishmanial tubulin inhibitors on L. donovani. Mol Biochem Parasitol 2000; 110:223-236.
32. Baum SG, Wittner M, Nadler JP. Taxol, a microtubule stablizing agent, blocks the replication of Trypanosoma cruzi. Proc Natl Acad Sci 1981; 78:4571-4575.
33. Moulay L, Robert-Gero M, Brown S et al. Sinefungin and taxol effects on cell cycle and cytoskeleton of Leishmania donovani promastigotes. Exp Cell Res 1996; 226:283-291.
34. Kapoor P, Sachdeva M, Madhubala R. Effect of the microtubule stablising agent taxol on leishmanial protozoan parasites in vitro. FEMS Microbiol Lett 1999; 176:429-435.
35. Dantas AP, Barbosa HS, De Castro SL. Biological and ultrastructural effects of the anti-microtubule agent taxol against Trypanosoma cruzi. J Submicrosc Cytol Pathol 2003; 35:287-294.
36. Luduena RF, Roach MC. Tubulin sulfhydryl groups as probes and targets for anitmitotic and antimicrotubule agents. Pharmacol Ther 1991; 49:133-152.
37. Rai S, Wolff J. Localization of the vinblastine-binding site on β-tubulin. J Biol Chem 1996; 271:14707-14711.
38. Chatterjee SK, Laffray J, Patel P et al. Interaction of tubulin with a new fluorescent analogue of vinblastine. Biochemistry 2002; 41:14010-14018.
39. Gigant B, Wang C, Ravelli RBG et al. Structural basis for the regulation of tubulin by vinblastine. Nature 2005; 435:519-522.
40. Grellier P, Sinou V, Garreau-de Loubresse N et al. Selective and reversible effects of vinca alkaloids on Trypanosoma cruzi epimastigote forms: Blockage of cytokinesis without inhibition of the organelle duplication. Cell Motil Cytoskeleton 1999; 42:36-47.
41. Ochola, DO, Prichard RK, Lubega GW. Classical ligands bind tubulin of trypanosomes and inhibit their growth in vitro. J Parasitol 2002; 88:600-604.
42. Floyd LJ, Barnes LD, Williams RF. Photoaffinity labelling of tubulin with (2-nitro-4-azidophenyl)deacetylcolchicine: Direct evidence for two colchicine binding sites. Biochemistry 1989; 28:8515-8525.
43. Uppuluri S, Knipling L, Sackett DL et al. Localization of the colchicine-binding site of tubulin. Proc Natl Acad Sci USA 1993; 90:11598-11602.
44. Bai R, Pei XF, Boyé O et al. Identification of cysteine 354 of β-tubulin as part of the binding site for the A ring of colchicine. J Biol Chem 1996; 271:12639-12645.
45. Ruoli B, Covell DG, Pei XF et al. Mapping the binding site of colchicinoids on β-tubulin. 2-Chloroacetyl-2-demethylthiocolchicine covalently reacts predominantly with cysteine 239 and secondary with cysteine 354. J Biol Chem 2000; 275:40443-40452.
46. Ravell RBG, Gigant B, Curmi PA et al. Insight into tubulin regulation from a complex with colchicine and a stathmin-like domain. Nature 2004; 428:198-202.
47. Shearwin KE, Timasheff SN. Effect of cochicine analogues on the dissociation of αβ tubulin into subunits: The locus of colchicine binding. Biochemistry 1994; 33:894-901.
48. Burns RG. Analysis of the colchicine-binding site of β-tubulin. FEBS Lett 1992; 297:205-208.
49. Macrae TH, Gull K. Purification and assembly in vitro of tubulin from Trypanosoma brucei brucei. Biochem J 1990; 265:87-93.
50. MacDonald LM, Armson A, Thompson RCA et al. Characterization of factors favouring the expression of soluble protozoan tubulin proteins in Escherichia coli. Protein Expr Purif 2003; 29:117-122.
51. Russell DG, Miller D, Gull K. Tubulin heterogeneity in the trypanosome Crithidia fasciculata. Mol Cell Biol 1984; 4:779-790.
52. Yakovich AJ, Ragone FL, Alfonzo JD et al. Leishmania tarentolae: Purification and characterization of tubulin and its suitability as a substrate for antileishmanial drug screening. Exp Parasitol 2006; in press.

53. Wenyon D. Observations on the intestinal protozoa of three Egyptian lizards, with a note on a cell-invading fungus. Parasitol 1921; 12:133-140.
54. Chan MMY, Fong D. Inhibition of leishmanias but not host macrophages by the antitubulin herbicide trifluralin. Science 1990; 249:924-926.
55. Traub-Cseko YM, Ramalho-Ortgão JM, Dantas AP et al. Dintroaniline herbicides against protozoan parasites; the case of Trpanosoma cruzi. Trends Parasitol 2001; 17:136-141.
56. Chan MMY, Grogl M, Chen CC et al. Herbicides to curb human parasitic infections: In vitro and in vivo effects of trifluralin on the trypanosomatid protozoans. Proc Natl Acad Sci USA 1993; 90:5657-5661.
57. Callahan HL, Kelley C, Pereira T et al. Microtubule inhibitors: Structure-activity analyses suggest rational models to identify potentially active compounds. Antimicrob Agents Chemother 1996; 40:947-952.
58. Morejohn LC, Fosket DE. The biochemistry of compounds with anti-microtubule activity in plant cells. Pharmacol Ther 1991; 51:217-230.
59. Bhattacharya G, Salem MM, Werbovetz KA. Anitleishmanial dinitroaniline sulfonamides with activity against parasite tubulin. Bioorg Med Chem Lett 2002; 21:2395-2398.
60. Bhattacharya G, Herman J, Delfín D et al. Synthesis and antitubulin activity of N^1- and N^4-substituted 3,5-dinitro sulfanilamides against African Trypanosomes and Leishmania. J Med Chem 2004; 47:1823-1832.
61. Werbovetz KA, Sackett DL, Delfín D et al. Selective antimicrotubule activity of N1-phenyl-3,5-dinitro-N4,N4-di-n-propylsulfanilamide (GB-II-5) against kinetoplastid parasites. Mol Pharmacol 2003; 64:1325-1333.
62. Wu D, George TG, Hurh E et al. Presystemic metabolism prevents in vivo antikinetoplastid activity of N1, N4-substituted 3,5-dinitro sulfanilamide, GB-II-150. Life Sci 2006; 79:1081-1093.
63. George TG, Johnsamuel J, Delfín DA et al. Antikinetoplastid antimitotic activity and metabolic stability of dinitroaniline sulfonamides and benzamides. Bioorg Med Chem 2006; 14:5699-5710.
64. Morrissette NS, Mitra A, Sept D et al. Dinitroanilines bind α-tubulin to disrupt microtubules. Mol Biol Cell 2004; 25:1960-1968.
65. Abraham I, Dion RL, Duanmu et al. 2,4-Dichlorobenzyl thiocyante, an antimitotic agent that alters microtubule morphology. Proc Natl Acad Sci USA 1986; 83:6839-6843.
66. Bai RL, Lin CM, Nguyen NY et al. Identification of the cysteine residue of β-tubulin alkylated by the antimitotic agent 2,4-dichlorobenzyl thiocyante, facilitated by separation of the protein subunits of tubulin by hydrophobic column chromatography. Biochem 1989; 28:5606-5612.
67. Szajnman SH, Yan W, Bailey BN et al. Design and synthesis of aryloxyethyl thiocyanate derivatives as potent inhibitors of Trypanosoma cruzi proliferation. J Med Chem 2000; 43:1826-1840.
68. Cottrell DM, Capers J, Salem MM et al. Antikinetoplastid activity of 3-aryl-5-thiocy-anatomethyl-1,2,4-oxadiazoles. Bioorg Med Chem 2004; 12:2815-2824.
69. Abraham I, Dion RL, Duanmu C et al. 2,4-Dichlorobenzyl thiocyanate, an antimitotic agent that alters microtubule morphology. Proc Natl Acad Sci USA 1986; 83:6839-6943.
70. Morgan RE, Westwood NJ. Screening and synthesis: High throughput technologies applied to parasitology. Parasitol 2004; 128:SS71-S79.
71. Baldwin J, Michnoff CH, Malmquist NA et al. High-throughput screening for potent and selective inhibitors of Plasmoduim falciparum dihydroorotate dehydrogenase. J Biol Chem 2005; 280:21847-21853.
72. St. George S, Bishop JV, Titus RG et al. Novel compounds active against Leishmania major. Antimicrob Agents Chemother 2006; 50:474-479.
73. Haggarty SJ, Mayer TU, Miyamoto et al. Dissecting cellular processes using small molecules: Identification of colchincine-like, taxol-like and other small molecules that perturb mitosis. Chem Biol 2000; 7:275-286.
74. Mayer TU, Kapoor TM, Haggarty SJ et al. Small molecule inhibitors of mitotic spindle bipolarity identified in a phenotype-based screen. Science 1999; 286:971-974.
75. Hamel E, Blokhin AV, Dale G et al. Limitations in the use of tubulin polymerization assays as a screen for the identification of new antimitotic agents: The potent marine natural product curacin A as an example. Drug Dev Res 1995; 34:110-120.
76. Kokoshka JM, Ireland CM, Barrows LR. Cell-based screen for identification of inhibitors of tubulin polymerization. J Nat Prod 1996; 59:1179-182.

CHAPTER 5

Fishing for Anti-Leishmania Drugs:
Principles and Problems

Emanuela Handman,* Lukasz Kedzierski, Alessandro D. Uboldi and James W. Goding

Abstract

To date, there are no vaccines against any of the major parasitic diseases including leishmaniasis, and chemotherapy is the main weapon in our arsenal. Current drugs are toxic and expensive, and are losing their effectiveness due to parasite resistance. The availability of the genome sequence of two species of *Leishmania*, *Leishmania major* and *Leishmania infantum*, as well as that of *Trypanosoma brucei* and *Trypanosoma cruzi* should provide a cornucopia of potential new drug targets. Their exploitation will require a multi-disciplinary approach that includes protein structure and function and high throughput screening of random and directed chemical libraries, followed by in vivo testing in animals and humans. We outline the opportunities that are made possible by recent technologies, and potential problems that need to be overcome.

Introduction

The 20th century saw huge progress with the discovery and application of antibiotics, which had a major impact on the burden of infectious diseases. By the late 1960s bacterial diseases seemed to be under control, and the USA Surgeon General was quoted as saying "We can now close the book on infectious disease". Antibiotics became a low priority in most pharmaceutical companies, and the emphasis shifted to drugs for chronic diseases.[1] The book is still open. Antibiotic resistance, newly emerging and reemerging diseases and the threat of bio-terrorism are leading to an increased realization that major public health risks are looming.[1-3]

The great success of antibiotics in the fight against bacterial infections suggested that a similar approach should be possible against other infectious diseases, but success against parasitic diseases has been limited. Much of the world's poorest populations suffer from one or more parasitic diseases, but in contrast to bacterial diseases, the pharmaceutical industry seems to have shown little interest in the search for antiparasitic drugs.[4]

As is the case with all parasitic diseases, there are no vaccines against leishmaniasis, and chemotherapy is the only means of treatment.[5-7] Disappointingly few drugs are available, and their efficacy is limited due to toxicity and increasing multiple drug resistance.[6,8,9] New drugs are needed, but their development will require new strategies, particularly in the choice of targets. Fortunately, recent technological advances are offering new opportunities for drug development. Moreover, there seems to be a welcome change in the attitude of the pharmaceutical industry away from a strictly profit-oriented approach to drug development and a greater pursuit of drugs against nonprofitable diseases.[10]

*Corresponding Author: Emanuela Handman—Walter and Eliza Hall Institute of Medical Research, Victoria, Australia. Email: handman@wehi.edu.au

Drug Targets in KinetoplastidParasites, edited by Hemanta K. Majumder.
©2008 Landes Bioscience and Springer Science+Business Media.

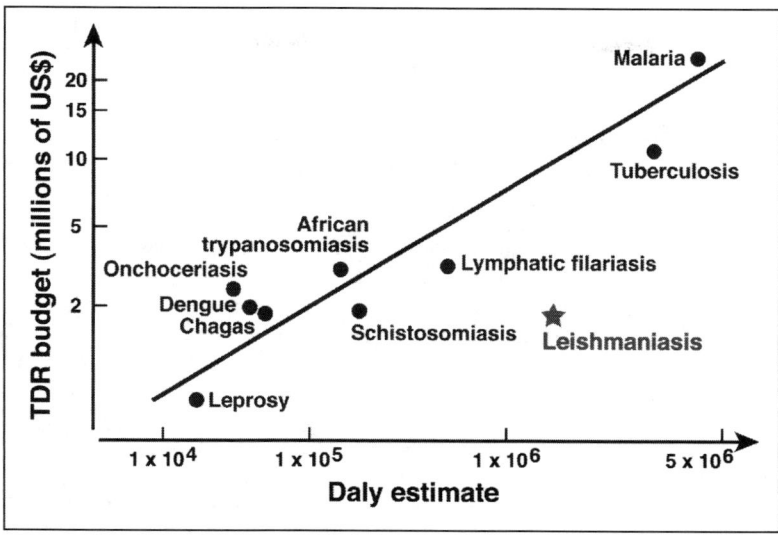

Figure 1. The DALY estimates determine the level of WHO/TDR funding for research; the higher the burden of disease, the higher the budget for its research and development. Modified from: Bergquist NR et al. Trends Parasitol 21(3):112-7;[24] ©2005 with permission from Elsevier.

The Disease Burden of Leishmaniasis

Leishmaniasis is a public health problem throughout most of the tropical and subtropical world,[4,11] and is a growing concern in war-torn countries such as Afghanistan,[12] Sudan[13] and Iraq.[14-16] It is caused by protozoan parasites of the genus *Leishmania*, and consists of a spectrum of diseases ranging in severity from a self-healing skin ulcer to fatal visceral disease. Leishmaniasis may be increasing its spread; it has been detected for the first time in East Timor[17] and in kangaroos in Australia.[18] Worldwide, there are 2 million new cases of symptomatic disease each year, and the incidence of infection is undoubtedly much greater if undiagnosed and subclinical infections are taken into account.[14,19] It has been estimated that one tenth of the world's population is at continuing risk of infection.[4,20-22]

In many regions leishmaniasis and HIV infections overlap. In Africa, South America and the Mediterranean basin, and immunosuppression from HIV often results in a change from controlled subclinical infection to one that is overwhelming and life threatening.[23] The World Health Organization has classified leishmaniasis as a category 1 disease, emerging and uncontrolled, reflecting the fact that control programs have largely failed.[5] The burden of leishmaniasis expressed in disability-adjusted life years (DALYs) is estimated to be over 2 million (ref. 14 and Fig. 1). Unfortunately, the large DALY and the WHO classification of disease severity have not been reflected in an appropriate budget allocation for leishmaniasis research.[24]

Transmission and Epidemiology

Like other kinetoplastids, *Leishmania* are transmitted by blood-feeding insects. Flagellated *Leishmania* promastigotes are transmitted to humans by the bite of sandfly vectors, and are taken up by receptor-mediated phagocytosis to become obligatory intracellular pathogens of macrophages and other phagocytic cells. In the macrophage, the parasites undergo a major differentiation program and transform into nonflagellated forms known as amastigotes, which are released from macrophages and spread the infection to other cells.

There are more than 20 *Leishmania* species that are pathogenic to humans, and about 70 sandfly species have been implicated as vectors which transmit the parasites.[5] There are two main epidemiological entities; the zoonotic type which includes animal reservoirs, and the anthroponotic type where humans are the only host. The *Leishmania* species have defined geographical ranges, which often reflect a combination of the parasite species and the presence of particular species of sand flies.[25,26]

Disease Manifestations

Although any *Leishmania* species is capable of causing more than one form of the disease, there is a strong tendency for particular species to cause a particular disease phenotype (Table 1).

The type and severity of leishmaniasis depends on the species and strain of the parasite, the genetics of the host and environmental factors.[27,28] Infection with parasites causing cutaneous leishmaniasis in the Old World such as *L. major* or *L. tropica* usually causes a cutaneous ulcer of 6-12 months duration. However, the disease severity varies considerably, with some *L. tropica*-infected individuals developing noncurable leishmaniasis recidiva in which there is progressive local tissue destruction.[29] In these individuals there are very few parasites in the affected tissues, and most of the tissue damage is caused by potent T cell-mediated host responses.[29] It is most likely that this form of disease reflects a particular host genotype that determines disease outcome. New World cutaneous leishmaniasis caused by the *L. braziliensis* complex can also lead to recurrent lesions, but these occur at sites distant from the initial site of infection.[30] In contrast, *L. donovani* and *L. infantum* usually cause visceral leishmaniasis, with no obvious lesion at the point of entry of the parasite. Infection with these organisms can remain asymptomatic or can cause massive splenomegaly and hepatomegaly. It is widely believed that the genotype of the host plays a major role in determining disease severity.[31-33]

Asymptomatic infections have also been described for the cutaneous leishmaniases, and these cases seem to be associated with the ability of the host to mount particularly effective antiparasite immune responses.[34] Nonetheless, these mild and largely asymptomatic cases do not seem to involve complete eradication of the parasite ("sterile immunity") because the

Table 1. *Geographical distribution and disease phenotypes caused by Leishmania*

Species	Clinical Manifestations	Geographical Distribution
L. mexicana complex	Cutaneous leishmaniasis and rare cases of mucocutaneous and diffuse cutaneous leishmaniasis	Central and northern parts of South America; rare cases in southern USA
L. braziliensis complex	Cutaneous leishmaniasis with some developing mucocutaneous leishmaniasis later	Central America and parts of South America
L. major	Cutaneous leishmaniasis	The Middle East, North and Central Africa, Southern Asia
L. tropica	Cutaneous leishmaniasis and rare cases of leishmaniasis recidiva	The Middle East and Southern Asia
L.aethiopica	Cutaneous leishmaniasis and rare cases of diffuse cutaneous leishmaniasis	Ethiopia
L. donovani	Visceral leishmaniasis with rare cases of post kala azar dermal leishmanoid	East Africa, sub Saharan Africa, Southern Asia including India
L. infantum	Cutaneous or visceral leishmaniasis depending on strain	Southern Europe, The Middle East, North Africa
L. chagasi	Visceral leishmaniasis and some atypical cutaneous leishmaniasis	Brazil, Venezuela and Colombia with isolated cases throughout South and Central America

parasites can often be detected if sufficiently sensitive techniques are used, and the disease may flare up in immunocompromised individuals.

Recovery from leishmaniasis is thought to depend on the induction of Th1 type immune responses and the production of a suite of pro- inflammatory cytokines.[5] However, the picture emerging from the study of the human response to infection is not as clear as that described for some mouse models of disease, where the induction of Th1 immune responses is necessary and sufficient for resistance to infection.[27]

Management and Control of Leishmaniasis

The main control strategy for leishmaniasis consists of diagnosis and treatment of infected individuals, vector control using pyrethroid insecticides or impregnated pyrethroid bednets, and control of the animal reservoir where feasible. Reservoir control performed with poisoned baits and culling of domestic dogs is very expensive and has not been particularly successful. Spraying with insecticides, often performed primarily for eradication of mosquitoes transmitting malaria has sometimes been successful, but the effect is usually transient,[14,35] which is hardly surprising given the enormous geographical spread of the insect vectors.

Treatment of leishmaniasis varies depending on the type of infection, the philosophy guiding the attending physicians, and the local availability of drugs. For cutaneous leishmaniasis, the options are not to treat and allow lesions to self-heal, to treat with a topically applied drug, or with an injectable drug. Visceral leishmaniasis must be treated systemically.[5,36] The chemotherapeutic arsenal is small. The front line drugs are pentavalent antimonials, sodium stibogluconate (Pentostam) and meglumine antimoniate (Glucantime). The second line drugs include pentamidine, azole derivatives and amphotericin B, but these are toxic and require long-term parenteral administration (refs. 6,7 and Fig. 2). Miltefosine is a new and orally administered drug, which was originally developed for the treatment of cancer and has been approved for use in visceral leishmaniasis in India.[5,6] Two new drugs that are currently in clinical trials are the aminoglycoside antibiotic paromomycin and the 8-aminoquinoline derivative sitamaquine.

Surprisingly, despite their use for over a quarter of a century, the mechanism of action and the identity of the actual biologically active molecules of most leishmaniasis drugs are not completely understood (for a recent review, see ref. 37). It does seem clear that, as is the case with most antibiotics, they work best in conjunction with the host's ability to mount appropriate immune responses. In leishmaniasis, the induction of IL-10, a Th2 type cytokine, has been associated with poor response to chemotherapy.[37,38] This may be an important consideration in the context of drug development, because some forms of leishmaniasis are associated with deleterious host responses.

In the 1970s there was much enthusiasm for the use of immunotherapy as an adjunct to chemotherapy, in particular for cutaneous and mucocutaneous leishmaniasis in South America.[39,40] This approach, while successful in some situations, has not been widely accepted or implemented.[5] An interesting development stemming from basic immunological studies on the role of the innate immune response to infection with *Leishmania*, has been the treatment of cutaneous leishmaniasis with an immunomodulating compound, imiquimod (Aldara™), which has been shown to act on dendritic cells and macrophages via the Toll like receptor-7, inducing secretion of pro-inflammatory cytokines that promote healing. Although not active on its own, imiquimod has led to a significant cure rate in patients resistant to other treatment when delivered in combination with meglumine antimoniate.[41]

A somewhat surprising observation has been the recent discovery that radiowave-induced heat treatment of cutaneous lesions can lead to the healing of treated as well as untreated lesions on different sites on the same individual.[42,43] The observation that the treatment induced the systemic production of pro-inflammatory cytokines suggested an immunological mechanism for the therapeutic effect.[42]

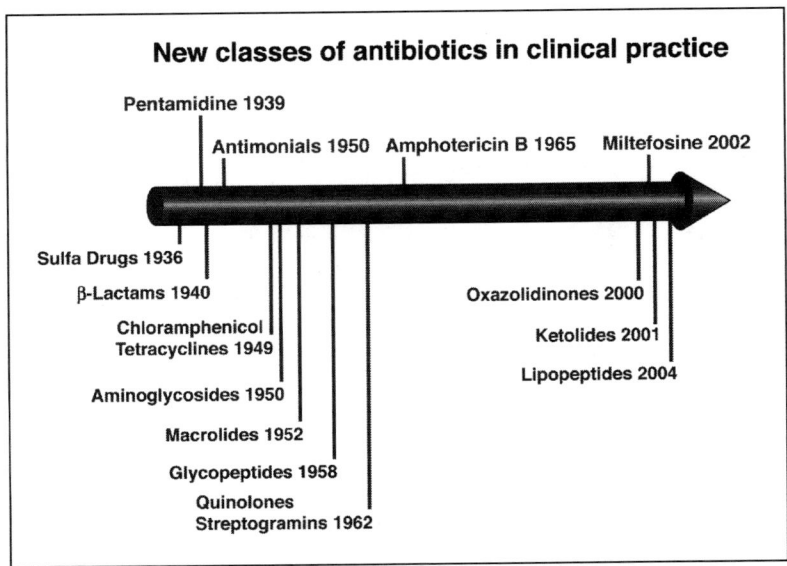

Figure 2. Existing antimicrobial and anti *Leishmania* drugs in clinical practice and the year of their development.

The Impact of New Technologies on Drug Discovery

The major factors that seem to drive the development of new drugs are advances in basic science, new technologies, and the dialogue between chemists and biologists.[44] Since the vast majority of drug targets are proteins, molecular biology has exerted a profound influence on drug discovery, initially through gene cloning and expression of potential drug targets.[44] However, the most important contribution of molecular biology to drug development may come from the analysis of the genome sequence of the important pathogens, which is revealing new points of attack based on novel enzymes and receptors. This knowledge, together with the relative ease of deleting and over-expressing genes, should lead to the identification of new drug targets. Knowledge of the complete sequence of the human genome will greatly facilitate this process by identifying genes that are unique to the parasite or sufficiently dissimilar to provide a basis for selective toxicity.

The other major factor that dominates the search for new drugs is the search for small molecules with therapeutic potential. Over the last few years, the pharmaceutical industry seems to have come to the realization that conventional strategies for finding new lead compounds for drug development have reached the limit of their effectiveness. In order to expand the diversity of potential libraries of chemical compounds, and to speed up the identification of new drug candidates, automation and "the magic of large numbers" are being incorporated into the pharmacological R&D programs.[44-46] This has resulted in an explosive increase in the number of new organic molecules available for bioassays. Coupled with new opportunities for target identification through genomics, proteomics and pharmacogenomics, high throughput enzyme and cell-based screening assays and large scale three dimensional structure determination of proteins, a new era for drug discovery is emerging.

High Throughput Screening

High throughput screening of large chemical libraries has become one of the most commonly used techniques for drug discovery in pharmaceutical research.[44,45] Automated robotic systems allow large numbers of potential drug targets to be tested in vitro for their effects on living cells or isolated cellular components such as enzymes or receptors. Tens or even hundreds of thousands of compounds can be tested in one day.

A complementary approach involves the use of more-focused libraries. Computational filtering methods have been employed to predict and eliminate molecules with unfavourable characteristics relating to absorption, distribution, metabolism and excretion (ADME), and other properties which may be associated with poor selectivity, promiscuous binding or poor uptake by cells.[45,46] To date, despite much promise, this approach has been somewhat disappointing, and has not eliminated the need for extensive empirical testing. However, it is possible that this will change in time with refinements of predictive algorithms and better selection and validation of the targets.[45]

Structure-Based Drug Discovery

Progress in structural biology both in NMR spectroscopy and X-ray crystallography and the huge increase in computing power have greatly facilitated the determination of the 3D structure of proteins. The feasibility of large-scale crystallization, at least for well-behaved soluble proteins, is now a reality as reflected in the rapid increase in the size of the database of available structures. The Structural Genomics of Pathogenic Protozoa (SGPP) consortium (http://depts.washington.edu/sgpp/) aims to express large numbers of proteins from a variety of parasites, including *Leishmania*, and to determine their three-dimensional crystal structures. These structures are available to all researchers and should provide targets for structure-based drug discovery. Unfortunately, this approach has mainly concentrated on the "low hanging fruit", because the majority of randomly chosen proteins do not crystallise easily and even if they do, the crystals often fail to diffract to sufficiently high resolution. Moreover, the crystallization of integral membrane proteins is still difficult and not amenable to the same style of "factory" approach used for soluble proteins. Since membrane proteins may comprise some 30% of all cellular proteins and include numerous potential drug targets, this is a serious bottleneck to progress.

In principle, the availability of the 3D structure of drug targets should enable the use of "virtual screening" of large databases of compounds in silico,[47] and promising candidates could then be tested for biological activity. In silico screening involves prediction of compound docking based on a target protein's structure, followed by searching for molecules which are similar to any compound identified in the screen. The computational docking approach was instrumental in the identification of 40 parasite-specific inhibitors of the *L. mexicana* glyceraldehyde-3-phosphate dehydrogenase (GAPDH).[48]

Proteomics

Proteomics may be defined as the attempt to analyse all proteins from a given source in terms of structure, function, expression, activity and interaction with other molecules. The hope has been that such a global approach may identify novel metabolic pathways and proteins which could be targeted by drugs. Global analysis at the protein level rather than gene transcription ("transcriptome") is particularly attractive for the identification of drug targets in *Leishmania* because the control of gene expression is post transcriptional and relies on the processing or stability of the mature mRNAs.

The most common proteomics approach uses two-dimensional gel electrophoresis to separate cellular proteins, followed by in-gel tryptic digestion of the protein spots and identification of the peptide sequences by mass spectrometry.[49,50] There are significant problems with this approach, which so far, have limited its usefulness for drug target discovery. A number of protein classes such as integral membrane proteins, positively charged and hydrophobic proteins are difficult to separate. There are often selective losses of individual proteins, for reasons that are not well understood, so the claim for a global picture is not entirely accurate. The approach is semi-quantitative at best. Importantly, low abundance proteins may not be detected.[51,52] However, newer approaches are constantly being developed.[53,54] Approaches which do not involve 2D gels may have the advantage of fewer selective losses, and the reduced resolution of 1D SDS gels may be compensated by additional chromatographic steps such as microbore reverse phase HPLC and capillary electrophoresis.[53,55]

The application of proteomics to *Leishmania* has generated some information on the promastigote stage of the parasite, but few data are available for the amastigote form.[56-58] Expression profiling of amastigotes has been hampered by the difficulty to obtain sufficient material of sufficient purity for analysis. This problem has been bypassed by the use of the so-called axenic amastigotes, which in some *Leishmania* species can be derived from promastigotes by culturing at 30-37°C and pH 5.0[59] although there is some doubt about how closely they mimic amastigotes produced from infected macrophages in vivo. The profile of the amastigote proteome is important because comparison of the transcriptomes of axenic and lesion-derived amastigotes has shown that axenic amastigotes are distinct from the lesion-derived amastigotes.[60] Comparison of the metabolites produced by the axenic and tissue-derived amastigotes also showed significant differences (McConville, personal communication).

The approaches described above involve denatured and fragmented proteins, and they aim to identify proteins based on electrophoretic mobility and amino acid sequence. A more directed approach aimed at inferring function for potential drug targets, involves the identification of protein-protein interactions. One approach has been the use of solid phase microarrays of proteins in their native conformation.[61,62] Disruption of such interactions would constitute a potential drug target. Identification of interactions between proteins can also be performed by the use of the yeast two-hybrid system, although this system is prone to false positives. Any putative interactions must be validated by some independent means (see below).[63,64] The yeast two-hybrid approach has recently culminated in putative comprehensive "interactome" maps.[64-67] However, detailed comparison of the interactions that were identified in putative "interactome" maps has shown a remarkable lack of agreement between different studies. The yeast two-hybrid system may also be limited by the fact that many protein-protein interactions in *Leishmania* may require unusual post-translational modifications which may not occur in yeast.

Monoclonal Antibodies

The application of human or "humanized" mouse monoclonal antibodies as therapeutic agents is steadily increasing, but it has not yet been applied to parasitic diseases. Although monoclonal antibodies can be targeted to very specific structures, they do not in general penetrate cell membranes, and their effect on *Leishmania* infection has not been tested sufficiently. Further disadvantages of monoclonal antibodies include the necessity to administer them by injection, extremely high cost, and the fact that even when "humanized" they may induce immune responses in their own right, including the possibility of life-threatening anaphylaxis.

Natural Products

The isolation and purification of the active ingredients of medicinal plants was one the major forces that led to the birth of the pharmaceutical industry in the 19th century.[44] After a long period of neglect, there is renewed interest in the analysis of "natural products" for the presence of inhibitors of enzymes essential for replication, cell cycle regulation or production of virulence factors.[68] The vast group of plant metabolites has already been shown to contain products with leishmanicidal activity.[69-71] The Drugs for Neglected Diseases Initiative (DNDi) is screening and investigating natural products (http://www.dndi.org/). Among the anti-*Leishmania* lead compounds are licochalcone A, extracted from the root of Chinese liquorice[72] and the propylquinolones chimanines A and D derived from the Bolivian plant *Galipea longiflora*.[68] So far, no mechanism of action has been described for any of the identified compounds and none has progressed from the laboratory to the clinic.[7]

Target Selection and Validation

It appears that the majority of current therapeutic targets have been enzymes or cell surface receptors because of the relative ease of modulating their function by small molecules.[44] The knowledge of the genome sequence of *Leishmania* and two other kinetoplastids, *T.brucei* and *T.cruzi* holds the promise of a rich source of new drug targets. How realistic is this promise, and what should we be looking for? The central aim of anti-microbials must be selective toxicity. To achieve this requires identification of structures and mechanisms that are either unique to the pathogen or sufficiently different to allow the development of drugs that are selective for the parasite, with minimal toxicity for the host.

Knowledge of individual enzymes and receptor-ligand interactions now provides the opportunity for a more directed approach to drug discovery and development. The "rational" mode of anti-microbial drug discovery has started to bear fruit with the identification of small molecule inhibitors of the anthrax lethal factor,[73] and the structure-guided discovery of a new inhibitor of RNA translation in bacteria.[74]

However, it must be acknowledged that many useful drugs have emerged from pragmatic screening without any concept or concern regarding their mechanism of action. Indeed, the mechanism of action of many useful drugs is still poorly understood.

Proteins and Metabolic Pathways Unique to the Parasite

Sterol biosynthesis in trypanosomatids is an attractive target because it differs from the mammalian host in that the final product is ergosterol rather than cholesterol.[75] One of the *Leishmania* targets already under investigation is the enzyme squalene synthase (SQS), which catalyzes the first step in this biosynthetic pathway by coupling two molecules of farnesylpyrophosphate to form squalene.[76]

The uniqueness of the trypanothione pathway in trypanosomatids has provided the opportunity to design highly selective anti-trypanosome drugs and also has promise for anti-*Leishmania* drugs.[77] Not only is trypanothione absent in humans, but it is essential for parasite survival because of its crucial role in regulating the cellular redox potential, and because the parasites do not have an alternative mechanism to protect against oxidative stress.[78] The enzymes that catalyse the last two steps in the biosynthesis of trypanothione are also important targets because they are absent in humans. These ATP-dependent C:N ligases catalyse the conjugation of the polyamine spermidine with two molecules of glutathione to generate trypanothione.[79]

The enzyme trypanothione reductase is another anti trypanosome drug target under current investigation, and represents a useful proof of concept for anti *Leishmania* drugs. Unlike the enzymes described above, it has a human counterpart, glutathione reductase, which has an amino acid sequence and structure similar to the parasite enzyme. However, the substantial differences in substrate specificity between the two enzymes has allowed the design of specific antiparasite inhibitors.[80] Among these are two natural products, cadabacine and lunarine.[80]

Another potential drug target involves the trafficking of proteins unique to the parasite into essential organelles, such as the mitochondrion/kinetoplast or glycosomes. In our laboratory, we have exploited the potential of mitochondrial proteins as drug targets. In trypanosomatids, the division of the kinetoplast is closely coordinated with the division of the flagellar basal body, the flagellum, and the cell itself.[81-83] The rationale behind this approach is that disruption of any part of this process might prevent cell division and lead to parasite killing.[84]

Since some of the targeting sequences which direct the import of cytosolic proteins into the *Leishmania* mitochondria are known and relatively conserved in kinetoplastids, we have used these sequences to identify a mitochondrial protein called miX (mitochondrial protein X) that occurs exclusively in *Leishmania* and *Trypanosomes*[85] and is absent from the mammalian genome. MiX is expressed throughout the *Leishmania* life cycle and appears to be essential for viability. Deletion of one allele of miX produces parasites with morphological and mitochondrial abnormalities, while deletion of both alleles seems to be lethal. In addition, the single allele knockout parasites display reduced infectivity in macrophages in vitro and reduced virulence in vivo, suggesting that miX may indeed be a good drug target.[85b]

Proteins Essential for Parasite Survival and Virulence Factors

Leishmania synthesize a variety of mannose-rich glycoconjugates which have been shown to be virulence factors. These glycoconjugates comprise glycoinositol phospholipid (GPI)-anchored glycoproteins, GPI-anchored lipophosphoglycan (LPG), free glycoinositolphospholipids (GIPLs) and proteophosphoglycans (PPG).[86] A prerequisite for the biosynthesis of glycoconjugates in *Leishmania*, as in all eukaryotes, is the conversion of monosaccharides to activated sugar nucleotides and dolicholphosphate derivatives. The activation of mannose comprises several enzymatic steps performed by phosphomannose isomerase (PMI), phosphomannomutase (PMM), GDP-mannose pyrophosphorylase (GDP-MP) and dolicholphosphate-mannose synthase (DPMS). The consecutive action of PMM and GDP-MP transform mannose-6-phosphate to GDP-mannose (GDP-Man), which is used as the mannose donor for all mannosylation reactions. Ilg and coworkers demonstrated that the deletion of the gene encoding PMM or GDP-MP, but not PMI or DPMS rendered the parasites avirulent, but still viable in culture.[87-90] Because of the importance of mannose-containing glycoconjugates to the parasite survival in the mammalian host, PMM and GDP-MP constitute attractive targets for anti *Leishmania* drug development.[90,91]

PMM (E.C. 5.4.2.8) is a phosphotransferase with a conserved phosphorylated motif DxDx(T/V).[92] The *L. mexicana* PMM is a member of the haloacid dehalogenase (HAD) family of proteins.[93-95] Our laboratory has solved the 3D structure of the *L. mexicana* PMM and shown that it has significant similarity to the human counterpart.[94,95] An interesting feature of the parasite PMM is the fact that it forms stable dimers.[94] It is not yet known whether dimerization is required for enzyme activity, but if it is, disruption of dimerization by small molecule inhibitors could disrupt its function. Since the mode of interaction between the subunits differs between the parasite and the human enzymes, this site may be an attractive drug target.

A second potential drug target in the mannose biosynthetic pathway is GDP-mannose pyrophosphorylase (GDP-MP) (E.C. 2.7.7.13), a member of the nucleotidyl transferase family of enzymes. Deletion of the gene in *Candida* and *S. cerevisiae* is lethal.[96] Like the PMM knockout parasites, the GDP-MP null parasites lack all mannose-containing glycoconjugates including the newly discovered β 1-2 mannan which is considered important in protection from stress.[90,91] There are two GDP-MP genes in humans, but little is known about their activity. No natural mutants have been observed in mammals including humans, suggesting that its loss is incompatible with life. Initial characterization of the leishmanial enzyme demonstrated that GDP-MP self-associates to form a stable hexamer.[97,98] Based on its biochemical characteriztion and structural similarities to other GDP-MPs, it has been proposed that the leishmanial hexamer is composed of a dimer of trimers, polymerisation being driven by noncovalent interactions between adjacent amino-terminal domains.[98] These interactions can be disrupted by relatively mild changes in pH and ionic strength, suggesting that it might be possible to design drugs that inhibit the assembly of the hexameric form of the enzyme.[97-99]

Target Validation

Any potential drug target must be expressed at the relevant stage in the parasite life cycle. For *Leishmania*, the relevant stage is the amastigote inside the macrophage. Assuming that this point can be demonstrated, the next step is to show that the modulation of expression of the target affects the survival of the parasite in macrophages in vitro and in animal models of disease. Most species of *Leishmania* that cause disease in humans also infect laboratory animals, but ultimate demonstration of usefulness requires clinical trials in humans.

The use of parasites with specific gene deletions provides a rapid and informative way to decide whether disrupting a pathway by drugs is likely to be an effective therapy. In *Leishmania*, this can be achieved by homologous recombination. However, complete knockout of gene function does not mimic the effects of drugs, which usually do not completely abrogate function. *Leishmania* are diploid, providing the opportunity to examine heterozygous "knockdown" parasites in which only one copy of the gene has been disrupted. To a first approximation, one may expect a 50% reduction in protein expression, although regulatory mechanisms may change this figure. Depending on the gene in question, a heterozygous loss can sometimes result in a clear phenotype (haploinsufficiency). This situation pertains to the miX protein in *Leishmania*, where heterozygous deletion results in morphologically abnormal parasites and reduced virulence, suggesting that even partial inhibition of this protein's function by drugs may have therapeutic effects.[85]

Unlike the situation in trypanosomes, RNAi does not seem to be operative in *Leishmania*. However, protein expression can be reduced by antisense RNA. This is usually achieved by transfection of parasites with an appropriate gene construct. Antisense RNA has been used to reduce the expression of the membrane protease gp63 and the amastigote-specific protein A2. Loss of either of the two proteins was shown to reduce parasite virulence, making them attractive drug targets.[100,101]

Another approach for validation of potential drug targets for *Leishmania*, which seems to be rich in multi copy genes that are difficult to delete, is the use of "dominant negative " mutants which may exert their effects by competing with the wild type protein for interacting partners or by inhibiting the formation of multimeric complexes. This approach led to the demonstration that the glycosomal protein GIM1 is essential for parasite survival.[102]

Concluding Remarks

The penultimate step in drug development is the testing of the drug candidate in animal models of disease, where all the complex interactions that underlie pathophysiological mechanisms take place. It is important that animal models not only manifest the relevant disease phenotype observed in humans, but that the underlying innate and adaptive immune responses are similar to the human disease. In the case of leishmaniasis, there are reasonably good animal models for some, but not all of the cutaneous forms. The animal models for visceral leishmaniasis are less satisfactory. Nonetheless, the animal models for leishmaniasis are considerably better than those for other parasitic diseases.

The final step in drug development inevitably involves clinical trials, with all their associated expense, logistical difficulties, risk and ethical problems. It is to be hoped that the powerful tools that are now at our disposal will result in a new generation of safer and more effective treatments in the near future.

Acknowledgements

We apologise to our many colleagues whose work we have not been able to cite due to space limitations. The authors are supported by the Australian Medical Research Council and UNDP/ World Bank/WHO/TDR Program.

References

1. Barrett JF. Can biotech deliver new antibiotics? Curr Opin Microbiol 2005; 8(5):498-503.
2. Greenfield RA, Bronze MS. Current therapy and the development of therapeutic options for the treatment of diseases due to bacterial agents of potential biowarfare and bioterrorism. Curr Opin Investig Drugs 2004; 5(2):135-40.
3. North CS, Pollio DE, Pfefferbaum B et al. Capitol hill staff workers' experiences of bioterrorism: Qualitative findings from focus groups. J Trauma Stress 2005; 18(1):79-88.
4. Hotez PJ, Molyneux DH, Fenwick A et al. Incorporating a rapid-impact package for neglected tropical diseases with programs for HIV/AIDS, Tuberculosis, and Malaria. PLoS Med 2006; 3(5):e102.
5. Murray HW, Berman JD, Davies CR et al. Advances in leishmaniasis. Lancet 2005; 366(9496):1561-77.
6. Croft SL, Barrett MP, Urbina JA. Chemotherapy of trypanosomiases and leishmaniasis. Trends Parasitol 2005; 21(11):508-12.
7. Davis AJ, Kedzierski L. Recent advances in antileishmanial drug development. Curr Opin Investig Drugs 2005; 6(2):163-9.
8. Ouellette M, Borst P. Drug resistance and P-glycoprotein gene amplification in the protozoan parasite Leishmania. Res Microbiol 1991; 142(6):737-46.
9. Croft SL, Coombs GH. Leishmaniasis—current chemotherapy and recent advances in the search for novel drugs. Trends Parasitol 2003; 19(11):502-8.
10. Herrling P. Experiments in social responsibility. Nature 2006; 439:267-268.
11. Ashford RW. The leishmaniases as emerging and reemerging zoonoses. Int J Parasitology 2000; 30:1269-1281.
12. Reithinger R, Mohsen M, Aadil K et al. Anthroponotic cutaneous leishmaniasis, Kabul, Afghanistan. Emerg Infect Dis 2003; 9(6):727-9.

13. Seaman J, Mercer AJ, Sondorp HE et al. Epidemic visceral leishmaniasis in southern Sudan: Treatment of severely debilitated patients under wartime conditions and with limited resources. Ann Intern Med 1996; 124(7):664-72.
14. Desjeux P. Leishmaniasis: Current situation and new perspectives. Comp Immunol Microbiol Infect Dis 2004; 27(5):305-18.
15. Korzeniewski K, Olszanski R. Leishmaniasis among soldiers of stabilization forces in Iraq. Review article. Int Marit Health 2004; 55(1-4):155-63.
16. Weina PJ, Neafie RC, Wortmann G et al. Old world leishmaniasis: An emerging infection among deployed US military and civilian workers. Clin Infect Dis 2004; 39(11):1674-80.
17. Chevalier B, Carmoi T, Sagui E et al. Report of the first cases of cutaneous leishmaniasis in East Timor. Clin Infect Dis 2000; 30:840.
18. Rose K, Curtis J, Baldwin T et al. Cutaneous leishmaniasis in red kangaroos: Isolation and characterization of the causative organisms. Int J Parasitol 2004; 34:655-664.
19. Desjeux P. Leishmaniasis. Nature Reviews Microbiology 2004; 2:692-693.
20. WHO/CTD. Leishmaniasis control. Burden and trends 1998:1-4.
21. Herwaldt BL. Leishmaniasis. Lancet 1999; 354:1191-1199.
22. Yardley V, Croft SL. A comparison of the activities of three amphotericin B lipid formulations against experimental visceral and cutaneous leishmaniasis. Int J Antimicrob Agents 2000; 13(4):243-8.
23. Laguna F. Treatment of leishmaniasis in HIV-positive patients. Ann Trop Med Parasitol 2003; 97(Suppl 1):135-42.
24. Bergquist NR, Leonardo LR, Mitchell GF. Vaccine-linked chemotherapy: Can schistosomiasis control benefit from an integrated approach? Trends Parasitol 2005; 21(3):112-7.
25. Bates PA, Rogers ME. New insights into the developmental biology and transmission mechanisms of Leishmania. Curr Mol Med 2004; 4(6):601-9.
26. Kamhawi S, Ramalho-Ortigao M, Pham VM et al. A role for insect galectins in parasite survival. Cell 2004; 119(3):329-41.
27. Handman E, Elso C, Foote S. Genes and susceptibility to leishmaniasis. Adv Parasitol 2005; 59:1-75.
28. Anderson CF, Mendez S, Sacks DL. Nonhealing infection despite Th1 polarization produced by a strain of Leishmania major in C57BL/6 mice. J Immunol 2005; 174:2934-2941.
29. Jacobson RL. Leishmania tropica (Kinetoplastida: Trypanosomatidae)—a perplexing parasite. Folia Parasitol (Praha) 2003; 50(4):241-50.
30. Bosque F, Saravia NG, Valderrama L et al. Distinct innate and acquired immune responses to Leishmania in putative susceptible and resistant human populations endemically exposed to L. (Viannia) panamensis infection. Scand J Immunol 2000; 51(5):533-41.
31. le Fichoux Y, Quaranta JF, Aufeuvre JP et al. Occurrence of Leishmania infantum parasitemia in asymptomatic blood donors living in an area of endemicity in southern France. J Clin Microbiol 1999; 37(6):1953-7.
32. Riera C, Fisa R, Udina M et al. Detection of Leishmania infantum cryptic infection in asymptomatic blood donors living in an endemic area (Eivissa, Balearic Islands, Spain) by different diagnostic methods. Trans R Soc Trop Med Hyg 2004; 98(2):102-10.
33. Blackwell JM, Mohamed HS, Ibrahim ME. Genetics and visceral leishmaniasis in the Sudan: Seeking a link. Trends Parasitol 2004; 20(6):268-74.
34. Follador I, Araujo C, Bacellar O et al. Epidemiologic and immunologic findings for the subclinical form of Leishmania braziliensis infection. Clin Infect Dis 2002; 34(11):E54-8.
35. Davies CR, Kaye P, Croft SL et al. Leishmaniasis: New approaches to disease control. Bmj 2003; 326(7385):377-82.
36. Blum J, Desjeux P, Schwartz E et al. Treatment of cutaneous leishmaniasis among travellers. J Antimicrob Chemother 2004; 53(2):158-166.
37. Bourreau E, Prevot G, Gardon J et al. High intralesional interleukin-10 messenger RNA expression in localized cutaneous leishmaniasis is associated with unresponsiveness to treatment. J Infect Dis 2001; 184(12):1628-30.
38. Murray HW. Prevention of relapse after chemotherapy in a chronic intracellular infection: Mechanisms in experimental visceral leishmaniasis. J Immunol 2005; 174(8):4916-23.
39. Convit J, Castellanos PL, Rondon A et al. Immunotherapy versus chemotherapy in localised cutaneous leishmaniasis. Lancet 1987; 1(8530):401-5.
40. Convit J, Castellanos PL, Ulrich M et al. Immunotherapy of localized, intermediate, and diffuse forms of American cutaneous leishmaniasis. J Infect Dis 1989; 160(1):104-115.
41. Arevalo I, Ward B, Miller R et al. Successful treatment of drug-resistant cutaneous leishmaniasis in humans by use of imiquimod, an immunomodulator. Clin Infect Dis 2001; 33(11):1847-51.
42. Reithinger R, Mohsen M, Wahid M et al. Efficacy of thermotherapy to treat cutaneous leishmaniasis caused by Leishmania tropica in Kabul, Afghanistan: A randomized, controlled trial. Clin Infect Dis 2005; 40(8):1148-55.

43. Reithinger R, Aadil K, Kolaczinski J et al. Social impact of leishmaniasis, Afghanistan. Emerg Infect Dis 2005; 11(4):634-6.
44. Drews J. Drug discovery: A historical perspective. Science 2000; 287(5460):1960-4.
45. Bleicher KH, Bohm HJ, Muller K et al. Hit and lead generation: Beyond high-throughput screening. Nat Rev Drug Discov 2003; 2(5):369-78.
46. Bleicher KH, Green LG, Martin RE et al. Ligand identification for G-protein-coupled receptors: A lead generation perspective. Curr Opin Chem Biol 2004; 8(3):287-96.
47. Bajorath J. Integration of virtual and high-throughput screening. Nat Rev Drug Discov 2002; 1(11):882-94.
48. Bressi JC, Verlinde CL, Aronov AM et al. Adenosine analogues as selective inhibitors of glyceraldehyde-3-phosphate dehydrogenase of Trypanosomatidae via structure-based drug design. J Med Chem 2001; 44(13):2080-93.
49. Aebersold R, Mann M. Mass spectrometry-based proteomics. Nature 2003; 422(6928):198-207.
50. Tyers M, Mann M. From genomics to proteomics. Nature 2003; 422(6928):193-7.
51. Huber LA. Is proteomics heading in the wrong direction? Nat Rev Mol Cell Biol 2003; 4(1):74-80.
52. Kopec KK, Bozyczko-Coyne D, Williams M. Target identification and validation in drug discovery: The role of proteomics. Biochem Pharmacol 2005; 69(8):1133-9.
53. Moritz RL, Clippingdale AB, Kapp EA et al. Application of 2-D free-flow electrophoresis/RP-HPLC for proteomic analysis of human plasma depleted of multi high-abundance proteins. Proteomics 2005; 5(13):3402-13.
54. Tang HY, Ali-Khan N, Echan LA et al. A novel four-dimensional strategy combining protein and peptide separation methods enables detection of low-abundance proteins in human plasma and serum proteomes. Proteomics 2005; 5(13):3329-42.
55. Simpson DC, Smith RD. Combining capillary electrophoresis with mass spectrometry for applications in proteomics. Electrophoresis 2005; 26(7-8):1291-305.
56. Gongora R, Acestor N, Quadroni M et al. Mapping the proteome of Leishmania Viannia parasites using two-dimensional polyacrylamide gel electrophoresis and associated technologies. Biomedica 2003; 23(2):153-60.
57. Drummelsmith J, Brochu V, Girard I et al. Proteome mapping of the protozoan parasite Leishmania and application to the study of drug targets and resistance mechanisms. Mol Cell Proteomics 2003; 2(3):146-55.
58. Drummelsmith J, Girard I, Trudel N et al. Differential protein expression analysis of Leishmania major reveals novel roles for methionine adenosyltransferase and S-adenosylmethionine in methotrexate resistance. J Biol Chem 2004; 279(32):33273-80.
59. Gupta N, Goyal N, Rastogi AK. In vitro cultivation and characterization of axenic amastigotes of Leishmania. Trends Parasitol 2001; 17(3):150-3.
60. Holzer TR, McMaster WR, Forney JD. Expression profiling by whole-genome interspecies microarray hybridization reveals differential gene expression in procyclic promastigotes, lesion-derived amastigotes, and axenic amastigotes in Leishmania mexicana. Mol Biochem Parasitol 2006; 146(2):198-218.
61. Templin MF, Stoll D, Schrenk M et al. Protein microarray technology. Drug Discov Today 2002; 7(15):815-22.
62. Delehanty JB, Ligler FS. Method for printing functional protein microarrays. Biotechniques 2003; 34(2):380-5.
63. Gietz ID. Yeast two-hybrid system screening. Methods Mol Biol 2005; 313:345-72.
64. Ito T, Ota K, Kubota H et al. Roles for the two-hybrid system in exploration of the yeast protein interactome. Mol Cell Proteomics 2002; 1(8):561-6.
65. LaCount DJ, Vignali M, Chettier R et al. A protein interaction network of the malaria parasite Plasmodium falciparum. Nature 2005; 438(7064):103-7.
66. Rual JF, Venkatesan K, Hao T et al. Towards a proteome-scale map of the human protein-protein interaction network. Nature 2005; 437(7062):1173-8.
67. Suthram S, Sittler T, Ideker T. The Plasmodium protein network diverges from those of other eukaryotes. Nature 2005; 438(7064):108-12.
68. Crump A. New medicines from nature's armamentarium. Trends Parasitol 2006; 22(2):51-4.
69. Fournet A, Munoz V. Natural products as trypanocidal, antileishmanial and antimalarial drugs. Curr Top Med Chem 2002; 2(11):1215-37.
70. Okpekon T, Yolou S, Gleye C et al. Antiparasitic activities of medicinal plants used in Ivory Coast. J Ethnopharmacol 2004; 90(1):91-7.
71. Takahashi M, Fuchino H, Sekita S et al. In vitro leishmanicidal activity of some scarce natural products. Phytother Res 2004; 18(7):573-8.
72. Zhai L, Chen M, Blom J et al. The antileishmanial activity of novel oxygenated chalcones and their mechanism of action. J Antimicrob Chemother 1999; 43(6):793-803.
73. Panchal RG, Hermone AR, Nguyen TL et al. Identification of small molecule inhibitors of anthrax lethal factor. Nat Struct Mol Biol 2004; 11(1):67-72.
74. Zhou Y, Gregor VE, Sun Z et al. Structure-guided discovery of novel aminoglycoside mimetics as antibacterial translation inhibitors. Antimicrob Agents Chemother 2005; 49(12):4942-9.

75. Urbina JA. Chemotherapy of Chagas disease. Curr Pharm Des 2002; 8(4):287-95.
76. Urbina JA, Concepcion JL, Rangel S et al. Squalene synthase as a chemotherapeutic target in Trypanosoma cruzi and Leishmania mexicana. Mol Biochem Parasitol 2002; 125(1-2):35-45.
77. Fairlamb AH. Chemotherapy of human African trypanosomiasis: Current and future prospects. Trends Parasitol 2003; 19(11):488-94.
78. Fairlamb AH. Metabolic pathway analysis in trypanosomes and malaria parasites. Philos Trans R Soc Lond B Biol Sci 2002; 357(1417):101-7.
79. Linares GE, Ravaschino EL, Rodriguez JB. Progresses in the field of drug design to combat tropical protozoan parasitic diseases. Curr Med Chem 2006; 13(3):335-60.
80. Hunter WN, Alphey MS, Bond CS et al. Targeting metabolic pathways in microbial pathogens: Oxidative stress and anti-folate drug resistance in trypanosomatids. Biochem Soc Trans 2003; 31(Pt 3):607-10.
81. Robinson DR, Gull K. Basal body movements as a mechanism for mitochondrial genome segregation in the trypanosome cell cycle. Nature 1991; 352(6337):731-3.
82. Robinson DR, Sherwin T, Ploubidou A et al. Microtubule polarity and dynamics in the control of organelle positioning, segregation, and cytokinesis in the trypanosome cell cycle. J Cell Biol 1995; 128(6):1163-1172.
83. Ogbadoyi EO, Robinson DR, Gull K. A high-order trans-membrane structural linkage is responsible for mitochondrial genome positioning and segregation by flagellar basal bodies in trypanosomes. Mol Biol Cell 2003; 14(5):1769-79.
84. Broadhead R, Dawe HR, Farr H et al. Flagellar motility is required for the viability of the bloodstream trypanosome. Nature 2006; 440(7081):224-7.
85. Uboldi AD, Walsh P, Spurck T et al. A Leishmania mitochondrial protein determines cell morphology, mitochondrial segregation and virulence. 2006.
85b. Uboldi AD, Lueder FB, Walsh P et al. A mitochondrial protein affects cell morphology, mitochondrial segregation and virulence in Leishmania. Int J Parasitol 2006; in press.
86. Ilgoutz SC, McConville MJ. Function and assembly of the Leishmania surface coat. Int J Parasitol 2001; 21:899-908.
87. Garami A, Ilg T. Disruption of mannose activation in Leishmania mexicana: GDP-mannose pyrophosphorylase is required for virulence, but not for viability. EMBO J 2001; 20(14):3657-3666.
88. Garami A, Ilg T. The role of phosphomannose isomerase in Leishmania mexicana glycoconjugate synthesis and virulence. J Biol Chem 2001; 276(9):6566-6575.
89. Garami A, Mehlert A, Ilg T. Glycosylation defects and virulence phenotypes of Leishmania mexicana phosphomannomutase and dolicholphosphate-mannose synthase gene deletion mutants. Mol Cell Biol 2001; 21(23):8168-8183.
90. Stewart J, Curtis J, Spurck TP et al. Characterisation of a Leishmania mexicana knockout lacking guanosine diphosphate-mannose pyrophosphorylase. Int J Parasitol 2005; 35(8):861-73.
91. Ralton JE, Nederer T, Piraino HL et al. Evidence that intracellular β1-2 mannan is a virulence factor in Leishmania parasites. J Biol Chem 2003; 278:40757-40763.
92. Collet JF, Stroobant V, Pirard M et al. A new class of phosphotransferases phosphorylated on an aspartate residue in an amino-terminal DXDX(T/V) motif. J Biol Chem 1998; 273(23):14107-12.
93. Koonin EV, Tatusov RL. Computer analysis of bacterial haloacid dehalogenases defines a large superfamily of hydrolases with diverse specificity. Application of an iterative approach to database search. J Mol Biol 1994; 244(1):125-32.
94. Kedzierski L, Malby RL, Smith BJ et al. Structure of Leishmania mexicana phosphomannomutase highlights similarities with human isoforms. J Mol Biol 2006; 363:215-27.
95. Silvaggi NR, Zhang C, Lu Z et al. The X-ray crystal structures of human alpha -phosphomannomutase 1 reveal the structural basis of carbohydrate deficient glycoprotein syndrome type 1a. J Biol Chem 2006.
96. Warit S, Zhang N, Short A et al. Glycosylation deficiency phenotypes resulting from depletion of GDP- mannose pyrophosphorylase in two yeast species. Mol Microbiol 2000; 36(5):1156-66.
97. Davis AJ, Perugini MA, Smith BJ et al. Properties of GDP-mannose pyrophosphorylase, a critical enzyme and drug target in Leishmania mexicana. J Biol Chem 2004; 279(13):12462-12468.
98. Perugini MA, Griffin MD, Smith BJ et al. Insight into the self-association of key enzymes from pathogenic species. Eur Biophys J 2005; 34(5):469-476.
99. Perez-Montfort R, Gomez-Puyou MT, Gomez-Puyou A. The interfaces of oligomeric proteins as targets for drug design against enzymes from parasites. Curr Top Med Chem 2002; 2(5):457-70.
100. Chen DQ, Kolli BK, Yadava N et al. Episomal expression of specific sense and antisense mRNAs in Leishmania amazonensis: Modulation of gp63 level in promastigotes and their infection of macrophages in vitro. Infect Immun 2000; 68(1):80-6.
101. Zhang WW, Matlashevski G. Loss of virulence in Leishmania donovani deficient in an amastigote specific protein A2. Proc Natl Acad Sci USA 1997; 94:8807-8811.
102. Flaspohler JA, Lemley K, Parsons M. A dominant negative mutation in the GIM1 gene of Leishmania donovani is responsible for defects in glycosomal protein localization. Mol Biochem Parasitol 1999; 99(1):117-28.

CHAPTER 6

Sterol 14-Demethylase Inhibitors for *Trypanosoma cruzi* Infections

Frederick S. Buckner*

Abstract

Chagas disease is caused by infection with the protozoan pathogen, *Trypanosoma cruzi*. The only approved therapeutics for treating Chagas disease are two nitroheterocyclic compounds (benznidazole and nifurtimox) that are suboptimal due to poor curative activity for chronic Chagas disease and high rates of adverse drug reactions. Sterol 14-demethylase inhibitors include azole antifungal drugs such as ketoconazole, fluconazole, itraconazole, and others. The first reports of potent activity of azole antifungal drugs against *Trypanosoma cruzi* came out about 25 years ago. Since then, a sizeable literature has accumulated on this topic. Newer triazole compounds such as posaconazole and D0870 have been shown to be effective at curing mice with chronic *Trypanosoma cruzi* infection. Small clinical studies with ketoconazole or itraconazole in humans with chronic Chagas disease have not demonstrated significant curative activity. However, there is good reason for optimism that newer compounds with greater potency and improved pharmacokinetic properties might be more efficacious. Data have been published demonstrating synergistic activity of azole drugs with various other compounds, indicating that combination chemotherapy may be an effective strategy as this field moves ahead. In light of the near absence of adequate therapeutics for curing patients with chronic Chagas disease, additional effort to develop better drugs needs to be a priority.

Introduction

In 1981, Docampo et al reported potent activity of the imadazole-containing antifungal drugs, miconazole and econazole, against *Trypanosoma cruzi* cultures.[1] Subsequently, considerable effort has gone into investigations of a variety of other imidazole or triazole compounds as potential therapeutics for Chagas disease. Most of these compounds, collectively referred to as azoles, have been derived from antifungal drug development programs. They act by inhibiting the sterol 14-demethylase enzyme, and they have been shown to block sterol biosynthesis in *Trypanosoma cruzi*. Some compounds are sufficiently active to cure mice of chronic *T. cruzi* infection. Unfortunately, this class of compounds has not been extensively studied in the clinical setting, and as a result the role of sterol 14-demethylase inhibitors for treatment of Chagas disease is not established. In this chapter, I will review biological basis for drug-targeting the sterol 14-demethylase enzyme of *T. cruzi*, and I will review the laboratory and clinical investigations of sterol 14-demethylase inhibitors for treatment of *T. cruzi* infections.

*Frederick S. Buckner—Department of Medicine, University of Washington, Seattle, Washington, USA. Email: fbuckner@u.washington.edu

Drug Targets in Kinetoplastid Parasites, edited by Hemanta K. Majumder.
©2008 Landes Bioscience and Springer Science+Business Media.

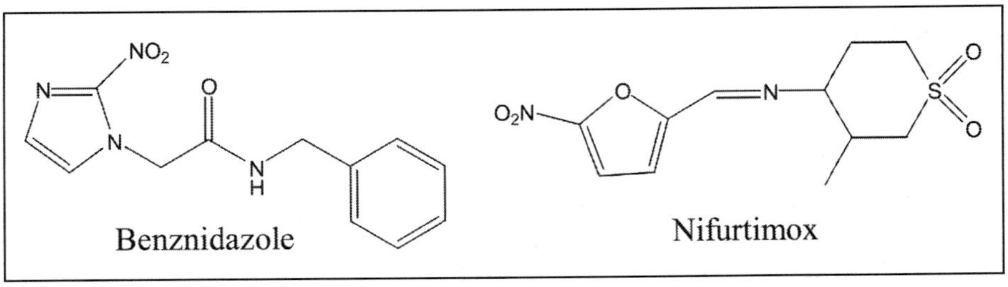

Figure 1. Chemical structures of nitroheterocycle compounds, benznidazole and nifurtimox, in clinical use for treatment of Chagas disease.

Chagas Disease

Also known as American trypanosomiasis, Chagas disease is endemic in 21 countries in the American hemisphere. Sixteen to 18 million people are infected, and 100 million people are at risk (http://www.who.int/ctd/chagas/burdens.htm). The etiologic agent, *T. cruzi*, is transmitted to humans primarily by blood feeding reduviid bugs that live in areas of poor housing or by blood transfusion. The epimastigote form replicates in the insect gut and transforms into infective metacyclic trypomastigotes. Trypomastigotes enter a variety of human cells, replicate intracellularly as round amastigotes, and transform to the motile trypomastigote form, which lyse the host cell. Trypomastigotes circulate, invade other host cells, or infect reduviid insects. Once infected with *T. cruzi*, the mammalian host remains infected for life. Chagas disease is manifested in three phases: acute, indeterminate, and chronic. The acute phase is the initial replicative phase for the organism in which demonstrable parasitemia and influenza-like symptoms occur. In the very young, direct invasion of parasites can cause fatal myocarditis or meningoencephalitis. The indeterminate phase is characterized by a lack of signs and symptoms. The chronic phase typically occurs 10-20 years after contracting the parasite and affects 10-30% of those infected.[2] It is manifested most commonly as debilitating or fatal cardiomyopathy or pathological dilations of the esophagus or colon. It is debated whether the pathogenesis of chronic Chagas disease is from inflammation directed to tissue parasites versus autoimmune phenomena.[3] Recent clinical evidence showed that aggressive antiparasitic therapy (using benznidazole) had a beneficial effect on cardiomyopathic progression,[4] suggesting an important role for etiologic treatment in the management of patients infected with *T. cruzi*.[5] The most widely used antiprotozoan treatments include benznidazole and nifurtimox (Fig. 1). In acute disease, nifurtimox results in parasitologic cure in about 70% of cases and decreases mortality. For children with indeterminant-phase of Chagas disease, benznidazole administered for 60 days has resulted in ~50% parasitologic cure.[6] Unfortunately, Chagas disease is usually diagnosed during the chronic phase when patients present with cardiac or GI symptoms, and treatment is less effective in this setting. One set of guidelines recommends etiologic treatment with benznidazole for 60 days for adults with chronic Chagas disease.[7] However, many practitioners are reluctant to use nifurtimox or benznidazole in chronic Chagas patients because of the unfavorable risk-to-benefit profile of these drugs, which are notorious for side effects such as vomiting and neuropathy.[5]

Drug Development for Chagas Disease

Any plans to make new therapeutics for Chagas disease need to bear in mind the characteristics of a drug that would allow it to succeed in clinical use. Essentially all the data in animal models and the clinical experience with humans indicate that long courses of treatment are required to produce cures. The situation is analogous to tuberculosis infection and systemic mycoses in which extended periods of drug pressure are necessary to kill off all the organisms.

Based on experience with benznidazole or nifurtimox, treatment courses of 8 weeks or longer will probably be necessary. A new drug should ideally have the following characteristics:

- Orally administered (any parenteral drug would be too difficult to give for long courses in resource limited settings)
- Once daily administration: this will greatly improve compliance and help insure continual drug pressure on tissue parasites for effectiveness and to reduce the chances of breeding resistant organisms.
- Low toxicity: clearly the major obstacle to delivering anti-Chagas therapy to infected populations has been intolerable toxicity of the benznidazole and nifurtimox
- Safe in children and women of reproductive age
- Low cost of goods: The drugs will mostly be used in poor populations in S. America.
- Stability: the drug will need to have a long shelf life in tropical temperatures

Sterol Biosynthesis—Overview

Azole antifungal compounds act by inhibiting the sterol 14-demethylase enzyme of the sterol biosynthetic pathway (Fig. 2). Sterols are essential lipid components of eukaryotic membranes. Sterols are important regulators of membrane permeability and fluidity. In addition, sterols have roles in aerobic metabolism, completion of cell cycle, sterol uptake, and

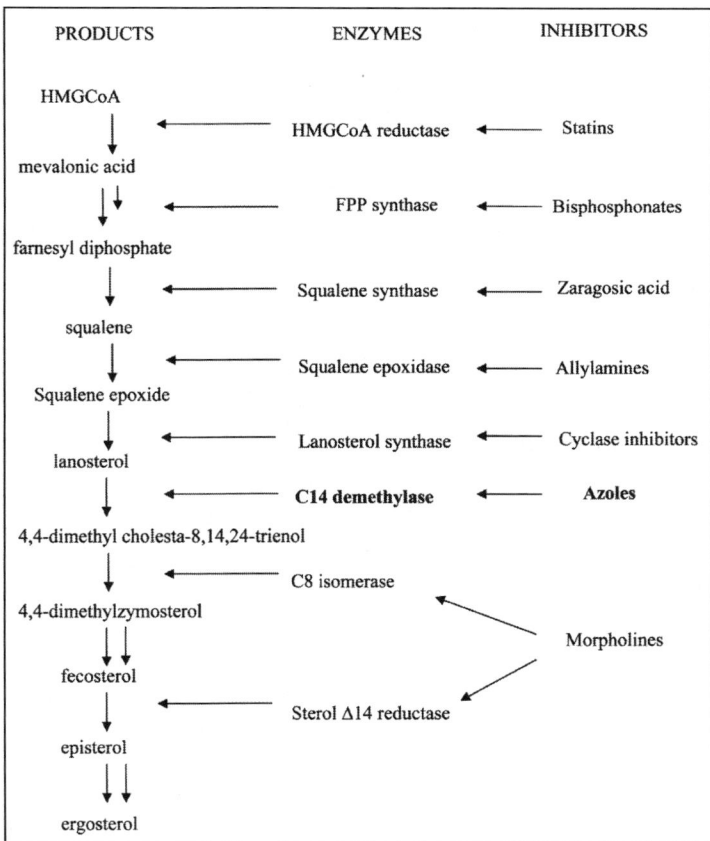

Figure 2. Sterol biosynthesis pathway as it is understood in yeast: biochemical products, enzymes, and inhibitors (double arrows represent more than one step).

sterol transport.[8] The final products of sterol biosynthesis vary among eukaryotes with mammals producing cholesterol, plants producing a variety of phytosterols, and fungi producing ergosterol.

The biosynthetic steps to produce ergosterol or cholesterol are the same in yeast and mammalian systems through the synthesis of 4,4-dimethylzymosterol (see Fig. 2). Despite identical reactions in the pre-zymosterol steps, differences in fungal and mammalian enzymes that catalyze these reactions have led to selectively active antifungals drugs such as azoles and terbinafine. Azole antifungal drugs have selective activity for the fungal sterol 14-demethylase enzyme over the mammalian ortholog.[9]

Many of the enzymes in the sterol biosynthesis pathway have selective inhibitors (Fig. 3). These include some extremely important classes of drugs such as "Statins", used to treat high cholesterol, bisphosphonates, used to treat osteoporosis, allylamines (e.g., terbinafine) and azoles used to treat fungal infections, and morpholines which are used as fungicidal agrochemicals.

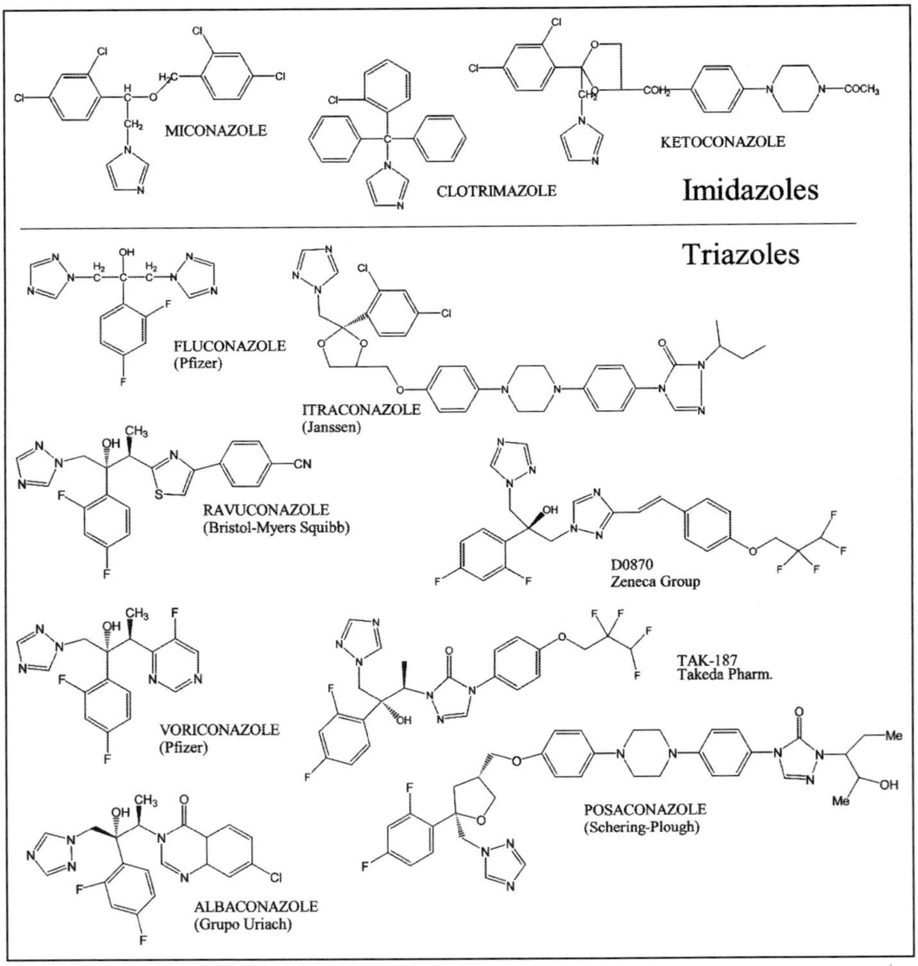

Figure 3. Structures of imidazole and triazole antifungal compounds under investigation or in clinical use.

Sterol 14-Demethylase (CYP51) and Azole Inhibitors

Synthesis of the major animal sterol (cholesterol), the fungal ergosterol, and the plant sito-sterol, requires removal of the 14α-methyl group from sterol precursors. The reaction is cata-lyzed by a microsomal cytochrome P450, the sterol 14-demethylase. In mammals and yeast, where the substrate is lanosterol, the enzyme is frequently called the lanosterol 14α-demethylase. The removal of the methyl group occurs through three successive oxidations resulting in decarbonylation, releasing formic acid (Fig. 4).[10]

Sterol 14-demethylase (CYP51) is a member of the cytochrome P450 superfamily of mixed-function oxidases that metabolize a structurally diverse group of exogenous and endog-enous organic substrates.[11] CYP51 is found in animals, plants, fungi, euglenozoa (e.g., trypanosomatids), and actinobacteria (e.g., Mycobacterium tuberculosis). Interestingly, the sterol 14-demethylase of *Trypanosoma brucei* (a closely related protist to *T. cruzi*) appears to use obtusifoliol as substrate rather than lanosterol.[12] The presence of the large phenylalanine side chain in the active site seems to lead to preferred processing of obtusifoliol.[13] The active site of *T. cruzi* con-tains an isoleucine in place of the Phe, consistent with lanosterol being the preferred substrate.[14]

Inhibition of cytochrome P450 enzymes by azole drugs results from coordination of the azole nitrogen to the heme iron, with the lipophilic ligand attached to the azole occupying the binding

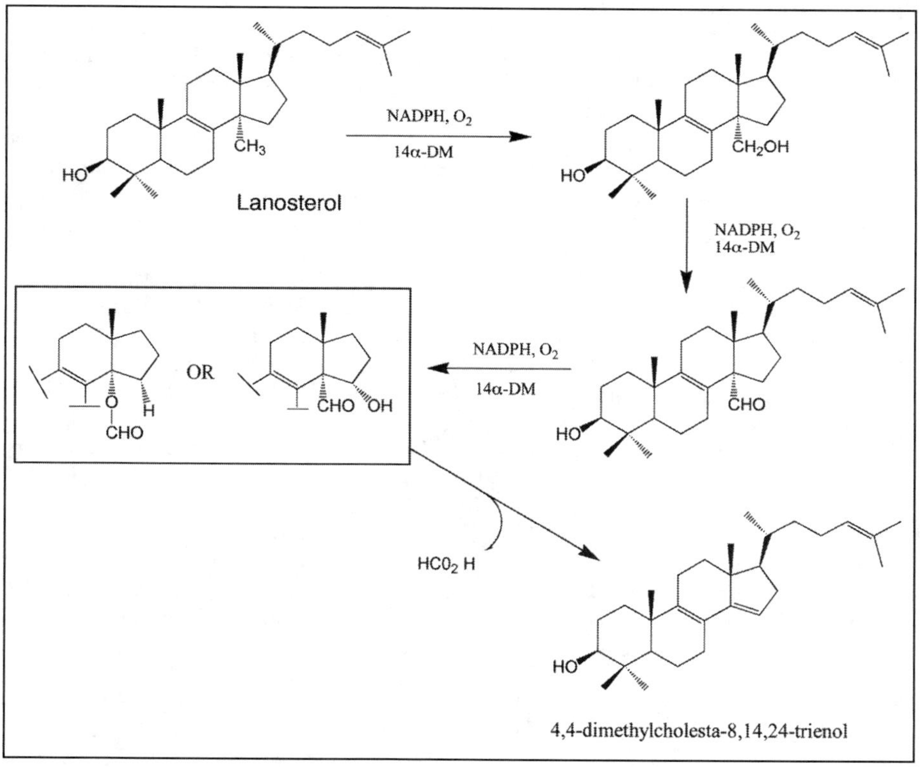

Figure 4. Mechanism of sterol 14-demethylase using lanosterol as substrate. This NADPH and O_2 dependent P450 enzyme oxidatively removes the 14-methyl group of lanosterol to form 4,4-dimethylcholesta-8,14,24 trienol. Adapted with permission from: Fischer RT, Trzaskos JM, Magolda RL et al. Lanosterol 14 alpha-methyl demethylase. Isolation and characterization of the third metabolically generated oxidative demethylation intermediate. J Biol Chem 1991; 266(10):6124-32.

site for lanosterol. These inhibitors prevent both binding of the substrate and oxygen activation.[15] During catalysis, the active-site heme iron is oxidized (from ferrous, Fe^{++}, to ferric, Fe^{+++}), and it is subsequently regenerated to its reduced form by a P450-reductase enzyme utilizing NADPH.

Sterol 14-demethylase (CYP51) is also present in humans, raising the obvious question of why sterol 14-demethylase inhibitors are not more toxic in mammals. As mentioned, azole antifungals (ketoconazole and itraconazole) have higher IC_{50}s against the human sterol 14-demethylase than against that of *Candida albicans* by a factor of ~30.[9] In addition, yeast cells have dramatically lower P450 enzyme content than mammalian cells. This combination makes yeast cells more sensitive to azole drugs.[16] One also needs to consider that cholesterol homeostasis in humans involves much more than just endogenous synthesis of cholesterol within each cell. The availability of cholesterol in the body is influenced by dietary intake and regulated absorption/excretion of cholesterol. This multidimensional system for maintaining cholesterol balance buffers the effect of inhibiting a single component of the system. Consequently, sterol biosynthesis inhibitors such as HMG-CoA reductase inhibitors (statins) or sterol 14-demethylase inhibitors (azoles) only partially lower the availability of cholesterol to cells in mammals, and thus are not highly toxic.

The main toxicities of azole compounds in humans are related to inhibition of P450 (CYP) enzymes. Humans have 57 sequenced CYP genes, and, like other mammals, contain CYP genes belonging to 15 families.[17] The functions of these CYP gene products primarily relate to metabolism of xenobiotics (and drugs) and to steroidogenesis. As can be surmised, inhibition of CYP enzymes is largely manifested as drug-drug interactions (by competition or induction of drug metabolizing enzymes) or as consequences of inhibiting steroid hormone biosynthesis. These effects limit the usefulness of ketoconazole, which is a promiscuous CYP inhibitor. Ketoconazole is notorious for potently inhibiting CYP3A4 and consequently causing elevated plasma concentrations of drugs metabolized by this important P450 enzyme. In addition, ketoconazole given to men at 200-400 mg/day transiently blocks testosterone and cortisol synthesis.[18-20] This is associated with hypogonadism in men and adrenal insufficiency in some cases. Ketoconazole also inhibits vitamin D metabolism in man.[20]

Newer generation azole antifungal drugs such as fluconazole and itraconazole (Fig. 3) have substantially few side effects in humans, and this is believed to be due to diminished activity against mammalian cytochrome P450 enzymes while retaining potent activity against fungal CYP51. These widely used and highly effective drugs establish a convincing precedent that it is possible to make selective inhibitors of microbial cytochrome P450 enzymes without causing dangerous cross-talk with mammalian cytochrome P450 enzymes.

Based on the large differences in amino acid sequences of the human and *Candida albicans* sterol 14-demethylase (39% sequence identity), it is not surprising that selective inhibitors for *Candida* were discovered. The *T. cruzi* sterol 14-demethylase protein is even more dissimilar to the human sterol 14-demethylase (31% identical) than is the *C. albicans* ortholog. It should also be pointed out that the *T. cruzi* and *C. albicans* CYP51 sequences are also very dissimilar (28% identity), suggesting that it is fortuitous that antifungal drugs are relatively potent against the *T. cruzi* CYP51. These data suggest there is an opportunity to develop drugs that are more specific for T. cruzi and that have greater potency against *T. cruzi* than existing antifungal azole drugs.

Sterols of *T. cruzi*

T. cruzi is most similar to fungi in its sterol composition, with ergosterol and ergosterol-like sterols being major membrane components.[21-23] The sterol biosynthesis pathway of *T. cruzi* presumably parallels that of yeast, although the full *T. cruzi* pathway has not been fully characterized. Clearly some differences exist late in the pathway due to the presence of sterol species with unusual alkyl variants attached at the C_{24} position. When grown with serum, approximately 25% of the membrane sterol is cholesterol.[24] Radiolabeled mevalonate is incorporated into the mature (4-desmethyl) sterols of *T. cruzi* epimastigotes that are grown with serum.[23] This demonstrates that de novo sterol biosynthesis is occurring. Azole antifungal drugs inhibit *T. cruzi* growth in vitro and alter the sterol composition of the cells.[1,24,25] Specifically, epimastigotes grown in the presence of ketoconazole or itraconazole have near complete suppression of the endogenous 14α-desmethyl sterols and accumulate 14α-methyl sterols.[24] These changes in membrane sterols are associated with profound ultrastructural alterations that lead to cell disintegration.[26]

The sterols of the replicative mammalian stage of *T. cruzi* (amastigotes) have also been described.[27] Exogenous sterol of host origin (cholesterol) account for approximately 80% of the weight of the total sterols in *T. cruzi* amastigotes.[27] Despite the fact the parasites incorporate sterols of host origin, they remain highly susceptible to sterol biosynthesis inhibitors. In fact, the antiproliferative potency of ketoconazole and other C-14 demethylase inhibitors is much greater against intracellular stages of *T. cruzi* compared with epimastigotes (see Table 1).[28-30] It has been suggested that this higher sensitivity is related to a smaller preformed sterol pool and/or a higher turn-over rate of sterols in these cells.[27] It also may be that a minimum amount of endogenous sterol with specific chemical characteristics is required for either structural purposes or for other reasons, such as cell cycle control.

Activity of Azole Compounds against *T. cruzi*

In Vitro

From published data, azole antifungal drugs are the most potent of all studied compounds with respect to in vitro potency against *T. cruzi* cultures. Several of these compounds have IC_{50} values in the picomolar range, specifically posaconazole, ravuconazole, and TAK-187 (Table 1). Since direct comparisons of IC50 can be difficult when assays are performed with different strains in different laboratories, we show data collected in our laboratory using the Tulahuen strain (Table 2). The compounds are ranked in terms of potency, demonstrating that posaconazole is the most potent and fluconazole is the least potent of those tested. The poor activity of fluconazole is noteworthy. Comparison of the structures suggests that the compounds with large sidegroups have greater activity. The potent in vitro activity of azole compounds is striking when compared to the clinically used drugs, benznidazole with an IC_{50} in the low micromolar range (Table 2). Despite the relatively weak in vitro activity of benzndizole, it is very active in animal models and remains the leading drug for treating humans infections. The reason(s) for the discordance between in vitro and in vivo activity of benznidazole is not understood. As already mentioned, the utility of benznidazole (and nifurtimox) is hindered by serious side effect profiles.

Enzyme Inhibition

Recombinant sterol 14-demethylase has been expressed in baculovirus (Buckner)[31] and more recently in *E. coli*.[32] The purified protein was shown to bind ketoconazole as evidenced by a shift in the aborbance spectrum of the enzyme.[31] A P450 enzyme type II binding spectrum was observed consistent with ketoconazole's imidazole moiety coordinating with the heme-iron in the active site. Similarly, a series of new synthetic bis-imidazole compounds with anti-*T. cruzi* activity were also shown to bind the recombinant enzyme.[33] The development of a high-throughput assay to screen compounds for binding affinity to the *T. cruzi* sterol 14-demethylase will enable investigators to broaden the search for more potent compounds.

Animal Studies

By far, the most widely used animal model for Chagas disease is murine infection with *T. cruzi*. Table 3 and 4 are summaries of the published data on the effectiveness of azole compounds for treating mice with *T. cruzi* infection. From these data, some general comments can be made:

- Mice with chronic *T. cruzi* infection are more difficult to cure using azole compounds than mice with acute infection.
- A subset of azole compounds (i.e., posaconazole, D0870, TAK-187) are more effective than others
- Azole compounds are not equally effective against all *T. cruzi* strains.
- Some azoles work synergistically with other drugs.

Table 1. *In vitro activity of sterol 14-demethylase inhibitors against T. cruzi (epimastigotes or intracellular amastigotes)*

Compound	Epimastigotes				Intracellular Amastigotes				
	Tc Strain	MIC (μM)	IC_{50} (μM)	Reference	Tc Strain	MIC (μM)	IC_{50} (μM)	Host Cell Type	Reference
Miconazole	Tulahuen		20	Docampo, 1981[1]					
Econazole	Tulahuen		20	Docampo, 1981[1]					
Fluconazole					Tulahuen		8.0	3T3 fibroblast	Buckner, 1996[48]
Itraconazole					Not stated	0.02	0.002	Perit. Macroph.	McCabe 1986[49]
Ketoconazole					Y		0.001	Perit. Macroph.	McCabe, 1984[50]
Ketoconazole	Peru		0.2	Beach, 1986[23]	Peru		0.01	VA-13 cells	Goad, 1989[24]
Ketoconazole	Y		0.1	Urbina, 1993[51]	Y		~0.002	Vero	Urbina, 1993[51]
D0870	Not stated		0.1	Urbina, 1996[35]	Not stated		0.01	Vero	Urbina, 1996[35]
Posaconazole	EP	0.02	0.014	Benaim, 2006[45]	EP	0.003	0.00025	Vero cells	Benaim, 2006[45]
Albaconazole	EP or Y	0.01		Urbina, 2000[52]	EP or Y	0.01	0.001	Vero	Urbina, 2000[52]
Ravuconazole	EP or Y	0.3	0.1	Urbina 2003a[37]	EP or Y	0.001	0.0001	Vero	Urbina 2003a[37]
TAK-187	EP or Y	0.3-1.0		Urbina 2003b[38]	EP or Y	0.001	0.0003	Vero	Urbina 2003b[38]

Table 2. In vitro activity of selected sterol 14-demethylase inhibitors (and benzndizole) against T. cruzi (Tulahuen strain) grown in murine 3T3 fibroblasts

	Intracellular Amastigotes IC_{50} (M)	Mouse Fibroblasts IC_{50} (M)
Posaconazole	0.0005	5
Ketoconazole	0.001	30
Itraconazole	0.001	1
Voriconazole	0.004	30
Miconazole	0.02	10
Benznidazole*	1.5	>20
Fluconazole	8	100

*Reference compound. From unpublished data, F.S. Buckner.

Azole compounds have been tested in the mouse model of Chagas disease in numerous laboratories. Direct comparisons between studies can be difficult because of the use of different protocols, different strains of *T. cruzi*, and different outcome measurements. A particularly problematic challenge with all of these studies is the test of cure of the animals. The tests include hemoculture, xenodiagnosis, serology, PCR, tissue analysis, and combinations of these. The reader is referred to the original papers for details.

Many of the azole compounds are able to cure mice with acute *T. cruzi* infection. These studies are generally performed by infecting mice with *T. cruzi* then initiating chemotherapy starting 24 hours post-infection. Compounds including ketoconazole, itraconazole, posaconazole, D0870, ravuconazole, and TAK-187 are able to cure mice acutely infected with *certain* strains of *T. cruzi*. Fluconazole did not cure acutely infected mice which is perhaps not surprising given its relatively poor in vitro activity against *T. cruzi* (Table 2). The poor activity of fluconazole is unfortunate since it is a widely used drug with a good safety record, and it is now a generic product.

The situation is different when compounds were used to treat mice with chronic *T. cruzi* infection. In general, "chronic" studies use mice that were infected at least 6-8 weeks before the initiation of treatment. Three methods were commonly used: (1) mice were infected with low enough numbers of parasites to allow most individuals to survive the initial parasitemic phase of infection, (2) mice were treated briefly with an anti-*T. cruzi* compound during the acute phase to prevent early mortality, or (3) certain *T. cruzi* strains were used that are able to establish chronic infection without causing much acute mortality. From these types of studies, it became evident that ketoconazole and itraconazole were unable to cure chronically infected mice.[34-36] Similarly, ravuconazole did not lead to parasitological cure of chronically infected mice.[37] It is thought that ravuconazole's relatively short serum half-life in mice (~4 hours) might be responsible for its poor activity in the chronic mouse model.[37]

The most impressive compounds for treating chronic *T. cruzi* infection are D0870 and posaconazole. D0870 cured 95% of mice chronically infected with the Bertoldo strain. Posaconazole cured 50-57% of mice chronically infected with CL, Y, or Colombiana strains. The studies on mice infected with the Columbiana strain are of note because this strain is benznidazole resistant. TAK-187 is another triazole compound with good activity against the Columbiana strain.[38] (TAK-187 was also shown to prevent cardiac damage in mice with *T. cruzi* infection[39]). These findings illustrate the important point that sterol 14-demethylase inhibitors have the potential to work against benznidazole-resistant strains. However, as discussed below, there are strains that resist both classes of compounds.

Table 3. Activity of sterol 14-demethylase inhibitors in mouse models of T. cruzi infection

Compound	Mouse Model*	Oral Dose	Duration of Rx	Tc Strain	% Survival	% Parasitological Cure	Reference
Ketoconazole	acute infxn	120-160 mg/kg/day	9 - 13 wks	Y		78-93%	McCabe, 1984[50]
Ketoconazole	acute infxn	15 mg/kg/day	14 days	Y	67%	33	Maldonado, 1993[44]
Ketoconazole	acute infxn	30 mg/kg/day	14 days	Y	100%	25%	Maldonado, 1993[44]
Ketoconazole	acute infxn	120 mg/kg/day	20	CL		93%	Brener, 1993[40]
Ketoconazole	acute infxn	120 mg/kg/day	20	Y		100%	Brener, 1993[40]
Ketoconazole	acute infxn	120 mg/kg/day	20	SC-28		80%	Brener, 1993[40]
Ketoconazole	acute infxn	120 mg/kg/day	20	Colombiana		8%	Brener, 1993[40]
Ketoconazole	acute infxn	120 mg/kg/day	20 days	CL		100%	Araujo, 2000[43]
Ketoconazole	established infxn	120 mg/kg/day	20 days	CL		0%	Araujo, 2000[43]
Ketoconazole	established infxn	120 mg/kg/day	20 days	Y		9%	Araujo, 2000[43]
Ketoconazole	established infxn	120 mg/kg/day	20 days	Colombiana		0%	Araujo, 2000[43]
Ketoconazole	acute infxn	20 mg/kg b.i.d.	28 + 15 days	Y	100%	20%	Urbina 2003a[37]
Ketoconazole	chronic infxn	60 mg/kg b.i.d.	18 wks	Y		0%	McCabe, 1988[34]
Ketoconazole	chronic infxn	80 mg/kg b.i.d.	7 wks	Y		0%	McCabe, 1988[34]
Ketoconazole	chronic infxn	30 mg/kg/day	21.5	Bertoldo	48%	4%	Urbina, 1996[35]
Ketoconazole	chronic infxn	30 mg/kg/day	28 + 15 days	Bertoldo	80%	0%	Urbina 2003a[37]
Itraconazole	acute infxn	15 or 30 mg/kg/day	7 days	Y, Tu, or CL	100%	0	McCabe, 1986[49]
Itraconazole	acute infxn	60 mg/kg/day	7-9 wks	Y	100%	<50%	McCabe, 1986[49]
Itraconazole	acute infxn	120 mg/kg/day	7-9 wks	Y	100%	100%	McCabe, 1986[49]
Itraconazole	chronic infxn	100 or 200 mg/kg/day	12 wks	Y		0%	Moreira, 1992[36]
Itraconazole	acute infxn	100 mg/kg/day	60 days	Genotype 32 (5 strains)		80%	Toledo, 2004[41]
Itraconazole	chronic infxn	100 mg/kg/day	60 days	Genotype 32 (5 strains)		69%	Toledo, 2004[41]
Itraconazole	acute infxn	100 mg/kg/day	60 days	Genotype 39 (5 strains)		54%	Toledo, 2004[41]
Itraconazole	chronic infxn	100 mg/kg/day	60 days	Genotype 39 (5 strains)		56%	Toledo, 2004[41]
Itraconazole	acute infxn	100 mg/kg/day	60 days	Genotype 19 (5 strains)		54%	Toledo, 2004[41]
Itraconazole	chronic infxn	100 mg/kg/day	60 days	Genotype 19 (5 strains)		44%	Toledo, 2004[41]
Itraconazole	acute infxn	100 mg/kg/day	60 days	Genotype 20 (5 strains)		0%	Toledo, 2004[41]
Itraconazole	chronic infxn	100 mg/kg/day	60 days	Genotype 20 (5 strains)		0%	Toledo, 2004[41]

continued on next page

Table 3. *Continued*

Compound	Mouse Model*	Oral Dose	Duration of Rx	Tc Strain	% Survival	% Parasito-Logical Cure	Reference
Fluconazole	acute infxn	200 mg/kg/day	4 wks	Y		0%	Campos, 1992[53]
Posaconazole	acute infxn	20 mg/kg/day	20 days	CL		100%	Molina 2000[54]
Posaconazole	acute infxn	20 mg/kg/day	20 days	Y		89%	Molina 2000[54]
Posaconazole	acute infxn	20 mg/kg/day	20 days	Colombiana		50%	Molina 2000[54]
Posaconazole	acute infxn	20 mg/kg/day	28 + 15 days	CL	90%	100%	Molina 2000[54]
Posaconazole	acute infxn	20 mg/kg/day	28 + 15 days	Y	90%	78%	Molina 2000[54]
Posaconazole	acute infxn	20 mg/kg/day	28 + 15 days	Colombiana	80%	75%	Molina 2000[54]
Posaconazole	acute infxn	20 mg/kg/day	28 + 15 days	SC-28	100%	100%	Molina 2000[54]
Posaconazole	acute infxn	20 mg/kg/day	28 + 15 days	VL-10	90%	56%	Molina 2000[54]
Posaconazole	chronic infxn	20 mg/kg/day	20 days	CL		57%	Molina 2000[54]
Posaconazole	chronic infxn	20 mg/kg/day	20 days	Y		50%	Molina 2000[54]
Posaconazole	chronic infxn	20 mg/kg/day	20 days	Colombiana		50%	Molina 2000[54]
D0870	acute infxn	20 mg/kg e.o.d.	56 days (28 doses)	Y	100%	65%	Urbina, 1996[35]
D0870	established infxn	20 mg/kg e.o.d.	56 days (28 doses)	Y	100%	60%	Urbina, 1996[35]
D0870	chronic infxn	20 mg/kg e.o.d.	56 days (28 doses)	Bertoldo	100%	95%	Urbina, 1996[35]
Ravuconazole	acute infxn	10 mg/kg b.i.d.	28 + 15 days	Y	100%	70%	Urbina 2003a[37]
Ravuconazole	acute infxn	20 mg/kg b.i.d.	28 + 15 days	Y	100%	40%	Urbina 2003a[37]
Ravuconazole	acute infxn	30 mg/kg b.i.d.	28 + 15 days	Y	100%	30%	Urbina 2003a[37]
Ravuconazole	acute infxn	15 mg/kg b.i.d.	20 days	Y	100%	58%	Urbina 2003a[37]
Ravuconazole	acute infxn	15 mg/kg b.i.d.	20 days	CL	100%	100%	Urbina 2003a[37]
Ravuconazole	acute infxn	15 mg/kg b.i.d.	20 days	Colombiana	100%	0%	Urbina 2003a[37]
Ravuconazole	chronic infxn	10 mg/kg b.i.d.	28 + 15 days	Bertoldo	100%	0%	Urbina 2003a[37]
Ravuconazole	chronic infxn	20 mg/kg b.i.d.	28 + 15 days	Bertoldo	90%	0%	Urbina 2003a[37]

continued on next page

Table 3. Continued

Compound	Mouse Model*	Oral Dose	Duration of Rx	Tc Strain	% Survival	% Parasito-Logical Cure	Reference
TAK-187	acute infxn	5 mg/kg/day	20 days	CL	100%	100%	Urbina 2003b[38]
TAK-187	acute infxn	10 mg/kg/day	20 days	CL	100%	100%	Urbina 2003b[38]
TAK-187	acute infxn	20 mg/kg/day	20 days	CL	100%	100%	Urbina 2003b[38]
TAK-187	acute infxn	20 mg/kg/e.o.d.	20 days	CL	100%	100%	Urbina 2003b[38]
TAK-187	acute infxn	5 mg/kg/day	20 days	Y	100%	60%	Urbina 2003b[38]
TAK-187	acute infxn	10 mg/kg/day	20 days	Y	100%	70%	Urbina 2003b[38]
TAK-187	acute infxn	20 mg/kg/day	20 days	Y	100%	70%	Urbina 2003b[38]
TAK-187	acute infxn	20 mg/kg/e.o.d.	20 days	Y	100%	70%	Urbina 2003b[38]
TAK-187	acute infxn	5 mg/kg/day	20 days	Colombiana	100%	30%	Urbina 2003b[38]
TAK-187	acute infxn	10 mg/kg/day	20 days	Colombiana	100%	50%	Urbina 2003b[38]
TAK-187	acute infxn	20 mg/kg/day	20 days	Colombiana	100%	60%	Urbina 2003b[38]
TAK-187	acute infxn	20 mg/kg/e.o.d.	20 days	Colombiana	100%	50%	Urbina 2003b[38]
TAK-187	acute infxn	20 mg/kg/e.o.d.	56 days (28 doses)	Y	100%	80%	Urbina 2003b[38]
Benznidazole	acute infxn	100 mg/kg/day	20 days	CL		93%	Brener, 1993[40]
Benznidazole	acute infxn	100 mg/kg/day	20 days	Y		100%	Brener, 1993[40]
Benznidazole	acute infxn	100 mg/kg/day	20 days	SC-28		4%	Brener, 1993[40]
Benznidazole	acute infxn	100 mg/kg/day	20 days	Colombiana		7%	Brener, 1993[40]
Benznidazole	established infxn	25 mg/kg/day	20 days	CL		0%	Araujo, 2000[43]
Benznidazole	established infxn	25 mg/kg/day	20 days	Y		0%	Araujo, 2000[43]
Benznidazole	established infxn	25 mg/kg/day	20 days	Colombiana		0%	Araujo, 2000[43]
Benznidazole	established infxn	50 mg/kg/day	20 days	CL		9%	Araujo, 2000[43]
Benznidazole	established infxn	50 mg/kg/day	20 days	Y		0%	Araujo, 2000[43]
Benznidazole	established infxn	50 mg/kg/day	20 days	Colombiana		0%	Araujo, 2000[43]
Benznidazole	established infxn	100 mg/kg/day	20 days	CL		100%	Araujo, 2000[43]
Benznidazole	established infxn	100 mg/kg/day	20 days	Y		31%	Araujo, 2000[43]
Benznidazole	established infxn	100 mg/kg/day	20 days	Colombiana		0%	Araujo, 2000[43]

continued on next page

Table 3. Continued

Compound	Mouse Model*	Oral Dose	Duration of Rx	Tc Strain	% Survival	% Parasito-Logical Cure	Reference
Benznidazole	acute infxn	100 mg/kg/day	20 days	CL	100%	100%	Urbina 2003a[37]
Benznidazole	acute infxn	100 mg/kg/day	20 days	Y	100%	75%	Urbina 2003a[37]
Benznidazole	acute infxn	100 mg/kg/day	20 days	Colombiana	100%	33%	Urbina 2003a[37]
Benznidazole	chronic infxn	100 mg/kg/day	20 days	CL		0%	Molina 2000[54]
Benznidazole	chronic infxn	100 mg/kg/day	20 days	Y		0%	Molina 2000[54]
Benznidazole	chronic infxn	100 mg/kg/day	20 days	Colombiana		0%	Molina 2000[54]
Nifurtimox	acute infxn	100 mg/kg/day	20 days	CL		93%	Brener, 1993[40]
Nifurtimox	acute infxn	100 mg/kg/day	20 days	Y		67%	Brener, 1993
Nifurtimox	acute infxn	100 mg/kg/day	20 days	SC-28		0%	Brener, 1993[40]
Nifurtimox	acute infxn	100 mg/kg/day	20 days	Colombiana		0%	Brener, 1993[40]
Nifurtimox	acute infxn	50 mg/kg/day	21.5 days	Y	80%	75%	Urbina 2003b[38]
Nifurtimox	acute infxn	50 mg/kg/day	30	Y	80%	50%	Benaim 2006[45]
Nifurtimox	chronic infxn	50 mg/kg/day	21.5	Bertoldo	60%	0%	Urbina, 1996[35]
Amiodarone	acute infxn	50 mg/kg e.o.d.	30 (15 doses)	Y	60%	0%	Benaim 2006[45]

*Acute infection: treatment usually started 24 h post-infection. *Established infection: treatment started after detection of patent parasitemia (7-15 days post-infection). *Chronic infection: treatment started after resolution of acute parasitemia phase (usually >45 days post-infection). *See indicated references for details.

Table 4. Combination treatment including sterol 14-demethylase inhibitors in the mouse models of T. cruzi infections

Compound	Mouse Model*	Oral Dose	Duration of Rx	Tc Strain	% Survival	% Parasito-Logical Cure	Reference
Ketoconazole + Lovastatin	acute infxn	40 mg/kg/day + 10 mg/kg/day	20 days	Y	80%	0%	Brener, 1993[40]
Ketoconazole + Terbinafine	acute infxn	15 mg/kg/day + 100 mg/kg/day	14 days	Y	100%	66-100%	Maldonado, 1993[44]
Ketoconazole + Benznidazole	chronic infxn	120 mg/kg/day + 50 mg/kg/day	20 days	CL		100	Araujo, 2000[43]
Ketoconazole + Benznidazole	chronic infxn	120 mg/kg/day + 50 mg/kg/day	20 days	Y		92.3	Araujo, 2000[43]
Ketoconazole + Benznidazole	chronic infxn	120 mg/kg/day + 50 mg/kg/day	20 days	Colombiana		0	Araujo, 2000[43]
Ketoconazole + Benznidazole	chronic infxn	120 mg/kg/day + 100 mg/kg/day	20 days	CL		100	Araujo, 2000[43]
Ketoconazole + Benznidazole	chronic infxn	120 mg/kg/day + 100 mg/kg/day	20 days	Y		100	Araujo, 2000[43]
Ketoconazole + Benznidazole	chronic infxn	120 mg/kg/day + 100 mg/kg/day	20 days	Colombiana		30.8	Araujo, 2000[43]
Posaconazole + amiodarone	acute infxn	20 mg/kg/day + 50 mg/kg e.o.d.	30 days	Y	100	80%	Benaim 2006[45]

Ketoconazole is active against the benznidazole-sensitive strain, CL, as well as against the benznidazole-resistant strains, SC-28 and YuYu. However, it is not active against the benznidazole-resistant strains, VL-10 or Colombiana.[40] Detailed studies with itraconazole further illustrate the differences in susceptibilities of different *T. cruzi* strains. A paper by Toledo et al[41] showed that *T. cruzi* strains of genotype 20 are widely resistant to itraconazole as well as benznidazole. These strains were collected from the Sao Paulo, Potosi, Sucre, and Cochabamba regions. It is interesting that these strains resist both itraconazole and benznidazole which do not share a common mechanism of action. The mechanism(s) of resistance to these drugs has not been established. The observation with genotype 20 strains illustrates the important point that variability in strain susceptibility is an important issue that cannot be ignored in the development of sterol 14-demethylase inhibitors, or, for that matter, any anti-*T. cruzi* drug.

A single large animal (dog) study has been performed investigating the activity of a sterol 14-demethylase inhibitor against *T. cruzi*.[42] Albaconazole (Fig. 3) was investigated in part because of its long in vivo half life in dogs and primates. Treatment was started 12-22 days post-infection, immediately after the appearance of parasitemia. Dogs were infected with the partially resistant Y strain and treated with albaconazole at 1.5 mg/kg/day for 60 or 90 days. All the treated animals survived (50% of controls died) and parasitemia resolved in 25% and 100% of dogs in the respective groups. Interestingly, dogs infected with the Berenice-78 strain (which is more sensitive to benznidazole than is the Y strain) did not do as well. When administered for 60, 90, or 150 days, none of the dogs was parasitologically cured. In contrast, benznidazole administered at 7 mg/kg b.i.d. for 60 days led to parasitologic cures in all dogs infected with either *T. cruzi* strain.

Some very important studies in mice have been performed looking at sterol 14-demethylase inhibitors in combination with other compounds (summarized in Table 4). Ketoconazole plus benznidazole led to high cure rates in mice chronically infected with the CL or Y strains.[43] This combination (in the higher dosing scheme) also led to a 31% cure rate against the resistant Colombiana strain. In the acute *T. cruzi* model, ketoconazole had synergistic activity with terbinafine (another sterol biosynthesis inhibitor acting on squalene epoxidase).[44] In contrast, lovastatin (an HMG-CoA reductase inhibitor) did not enhance the benefitial effect of ketoconazole in the acute mouse model of *T. cruzi* infection.[40] Finally, posaconazole was used in combination with the anti-arrhythmic drug, amiodarone.[45] The authors showed that amiodarone has its own intrinsic activity against *T. cruzi* by interfering with parasite calcium homeostasis and by blocking ergosterol biosynthesis. Posaconazole and amiodarone worked synergistically in vitro against *T. cruzi*. In the acute infection model using groups of 10 mice, posaconazole alone resulted in 60% cures, amiodarone alone resulted in 0% cures, and posaconazole + amiodarone resulted in 80% cures.

Clinical Studies

Only two azole antifungal drugs have been investigated in humans for treatment of Chagas disease: ketoconazole, and itraconazole. Ketoconazole was given to 8 patients with chronic Chagas disease for 10-14 weeks with no evidence of parasitologic cure as measured by hemoculture and serology.[40] The ketoconazole was given orally at the standard dose range (4.5 -8.8 mg/kg/day). The patients were Brazilian from the Minas Gerais state. The strains of the infecting parasites were not reported.

Two studies using itraconazole have been reported. One was performed by investigators at University of Sao Paulo, Brazil, involving 18 patients with chronic Chagas disease. Patients were given itraconazole at 100 mg/day or 200 mg/day for 12 weeks. There were no parasitologic cures.[36]

The other, larger study on itraconazole was led by investigators in Santiago, Chile.[46] Patients (ages 9-50) with chronic Chagas disease were recruited from urban and rural parts of Chile and Argentina. A total of 505 patients were divided into three treatment groups receiving itraconazole (400 mg/day for 120 days) or allopurinol (8.5 mg/kg/day for 60 days) or placebo. Adverse drug events in the itraconazole or allopurinol groups did not differ significantly from the placebo group. Follow-up at 4 years showed no changes in antibody titers using conventional methods (IF, ELISA, or Western blots), although 14 patients in the treatment groups had decreased antibody titers by IHA (9 treated with allopurinol, 5 treated with itraconazole, 0 treated with placebo). Xenodiagnosis was used to help establish parasitological cure. The analysis was complicated by the fact that large numbers (400) of patients had negative xenodiagnostic cultures at the beginning of the study. However, of the patients with baseline positive xenodiagnostic cultures, a significantly larger number converted to negative in the itraconazole group (38 of 43, 88%) or the allopurinol group (18 of 29, 62%) compared to the placebo group (4 of 16, 25%).

The patients were followed up 11-years post-treatment. Only patients treated with itraconazole or allopurinol were included in this analysis (apparently because most of the placebo-treated patients were subsequently treated with one of the drugs). The follow-up study focused on a subset of 109 patients from a region without ongoing *T. cruzi* transmission.[47] Only 21 (19.3%) of patients were negative in all the parasitologic tests and all patients remained seropositive by IFAT and/or ELISA. Thus, it does not appear that the treatment with itraconazole (or allopurinol) led to significant numbers of parasitologic cures in this patient population. An analysis of ECGs from baseline and at 11-years indicated that a substantial number of patients had ECGs that remained normal over time, or changed from abnormal to normal (51% in itraconazole-treated patients and 67% in allopurinol treated). Without a comparison to a placebo group, however, it is not possible to confidently say that these findings were due to therapeutic benefit of the treatments.

The current clinical studies have not established a clear role for using either ketoconazole or itraconazole for treatment of patients with Chagas disease, although it is possible that there may be a marginal therapeutic benefit with itraconazole. Based on the encouraging data with newer azole drugs in animal models (clearly with superior activity than ketoconazole or itraconazole), one has to be hopeful that additional clinical studies will be performed with more active sterol 14-demethylase inhibitors in the near future.

The Development Status of Sterol 14-Demethylase Inhibitors

Of the compounds described in Tables 1-4 and Figure 3, only ketoconazole, fluconazole, itraconazole, and voriconazole are FDA approved for clinical use (as of March, 2006). These drugs are all approved as antifungal agents. As was discussed, ketoconazole and itraconazole have been subjected to limited studies in patients with Chagas disease. To my knowledge, no published clinical research on Chagas disease with either fluconazole or voriconazole is available. Since fluconazole is relatively inactive against *T. cruzi*, a clinical study with this drug is probably not warranted. D0870 was found to be associated with cardiotoxic side effects during clinical studies for application as an antifungal drug, and its development was discontinued. The remaining compounds including posaconazole, ravuconazole, TAK-187, and albaconazole are in various stages of clinical or preclinical development as antifungal drugs. If they are approved for human use, they may become candidates for a second-use application for Chagas disease. It remains to be seen whether a business entity or another funding agency would sponsor the necessary clinical trial(s) to establish the utility of one or more of these compounds for Chagas disease.

Can a Better Sterol 14-Demethylase Inhibitor Be Developed against *T. cruzi*?

All of the compounds shown in Figure 3 came from pharmaceutical research programs to develop antifungal drugs. These compounds were developed and optimized for activity against fungi. Consequently, the use of these compounds for treating *T. cruzi* infections is a second-use application. There are obvious practical and economic reasons for this approach. However, it cannot be ignored that the investigators dedicated to finding therapeutics for Chagas disease are largely dependent on luck whether or not these antifungal compounds will also have activity against *T. cruzi*. As was discussed above, it is very fortunate that so many antifungal compounds do have good *T. cruzi* activity knowing that the overall amino acid sequence identity of the sterol 14-demethylase enzyme of *Candida albicans* and *T. cruzi* is only 28%. The substantial differences of the enzymes strongly suggests that a directed effort to develop specific anti-*T. cruzi* sterol 14-demethylase inhibitors might yield even more potent and biologically active compounds than those that have been discovered by "piggy-backing" onto antifungal drug development programs. Two new classes of imidazole-containing compounds have been reported recently that could potentially serve as starting points for medicinal chemistry efforts to develop inhibitors specifically tailored to inhibit the *T. cruzi* sterol 14-demethylase. These include a series of disubstituted imidazoles[33] and derivatives of the anticancer drug, tipifarnib.[14] The authors report a structural model of the *T. cruzi* sterol 14-demethylase enzyme based on the known structure of a prokaryotic CYP51 ortholog,[14] and propose a rational drug design program to synthesize compounds with optimized activity against the *T. cruzi* sterol 14-demethylase.

Summary

- As monotherapy in mice, the best sterol 14-demethylase inhibitors appear to be at least as effective as benznidazole or nifurtimox for treating *T. cruzi* infections. Three azole compounds deserve special notice are posaconazole, ravuconazole, and TAK-187. A fourth compound, D0870 had excellent activity, but was discontinued from clinical development due to serious side effects in clinical trials for fungal infections.
- The triazole antifungals are generally better tolerated and safer than the nitroheterocycle compounds (Fig. 1) which are notorious for severe side-effects
- Sterol 14-demethylase inhibitors have not been extensively investigated for treatment of human Chagas disease, thus it is not yet possible to judge whether this chemical class clearly has or will have a role in treating human cases. It is evident that prolonged courses of therapy with drugs such as itraconazole are much better tolerated than treatments with nitroheterocycle compounds.
- Sterol 14-demethylase inhibitors with more potent activity and/or more optimal pharmacokinetic properties than ketoconazole and itraconazole need to be investigated in clinical trials to evaluate their potential for treating humans with Chagas disease
- The use of sterol 14-demethylase inhibitors in combination with other compounds needs to be more extensively investigated for the following reasons:
  - No single drug (including azole compounds) may have sufficient activity to effect parasitologic cure in chronically infected humans, thus combination therapy may be absolutely necessary
  - Synergy studies in mice with sterol 14-demethylase inhibitors and other classes of compounds (e.g., terbenifine, benznidazole, amiodarone) indicate that synergistic or additive effects are possible
  - Combination therapy could potentially shorten the course of therapy necessary to effect cure
  - Combination therapy could potentially allow for the use of lower doses of drugs with reduced side effects
  - Combination therapy could potentially reduce the risk of promoting resistant strains of *T. cruzi*
- *New* sterol 14-demethylase inhibitors that are specifically designed for activity against *T. cruzi* may open the opportunity for even more potent therapeutics than those compounds obtained from antifungal drug development programs

References

1. Docampo R, Moreno SN, Turrens JF et al. Biochemical and ultrastructural alterations produced by miconazole and econazole in Trypanosoma cruzi. Mol Biochem Parasitol 1981; 3:169-80.
2. World Health Organization. Control of Chagas' disease. WHO Tech Rep Ser 1991; 811:1-93.
3. Eisen H, Petry K, Van Voorhis WC. The origin of autoimmune pathology associated with Trypanosoma cruzi infection. In: Van Der Ploeg LHT, Cantor CR, Voegel HJ, eds. Immune Recognition and Evasion: Molecular Aspects of Host-Parasite Interaction. London: Academic Press Inc., 1990:91-103.
4. Viotti R, Vigliano C, Armenti H et al. Treatment of chronic Chagas' disease with benznidazole: Clinical and serological evolution of patients with long-term follow-up. Am Heart J 1994; 127:151-62.
5. Urbina JA, Docampo R. Specific chemotherapy of Chagas disease: Controversies and advances. Trends Parasitol 2003; 19:495-501.
6. Sosa ES, Segura EL, Ruiz AM et al. Efficacy of chemotherapy with benznidazole in children in the indeterminate phase of Chagas' disease. Am J Trop Med Hyg 1998; 59:526-29.
7. Pan American Health Organization/World Health Organization. Etiological treatment for Chagas disease. Rev Pat Trop 1998; 28:247-79.
8. Daum G, Lees ND, Bard M et al. Biochemistry, cell biology and molecular biology of lipids of Saccharomyces cerevisiae. Yeast 1998; 14:1471-510.
9. Lamb DC, Kelly DE, Waterman MR et al. Characteristics of the heterologously expressed human lanosterol 14alpha-demethylase (other names: P45014DM, CYP51, P45051) and inhibition of the purified human and Candida albicans CYP51 with azole antifungal agents. Yeast 1999; 15:755-63.
10. Fischer RT, Stam SH, Johnson PR et al. Mechanistic studies of lanosterol 14 alpha-methyl demethylase: Substrate requirements for the component reactions catalyzed by a single cytochrome P-450 isozyme. J Lipid Res 1989; 30:1621-32.
11. Rozman D, Stromstedt M, Waterman MR. The three human cytochrome P450 lanosterol 14 alpha-demethylase (CYP51) genes reside on chromosomes 3, 7, and 13: Structure of the two retrotransposed pseudogenes, association with a line-1 element, and evolution of the human CYP51 family. Arch Biochem Biophys. 1996; 333:466-74.
12. Lepesheva GI, Nes WD, Zhou W et al. CYP51 from Trypanosoma brucei is obtusifoliol-specific. Biochem 2004; 43:10789-99.
13. Podust LM, Yermalitskaya LV, Lepesheva GI et al. Estriol bound and ligand-free structures of sterol 14alpha-demethylase. Structure (Camb.) 2004; 12:1937-45.
14. Hucke O, Gelb MH, Verlinde CL et al. The protein farnesyltransferase inhibitor Tipifarnib as a new lead for the development of drugs against Chagas disease. J Med Chem 2005; 48:5415-18.
15. Walker KA, Kertesz DJ, Rotstein DM et al. Selective inhibition of mammalian lanosterol 14 alpha-demethylase: A possible strategy for cholesterol lowering. J Med Chem 1993; 36:2235-37.
16. Vanden Bossche H, Marichal P, Gorrens J et al. Interaction of azole derivatives with cytochrome P-450 isozymes in yeast, fungi, plants and mammalian cells. Pestic Sci 1987; 21:289-306.
17. Lewis DF. 57 varieties: The human cytochromes P450. Pharmacogenomics 2004; 5:305-18.
18. Pont A, Williams PL, Azhar S et al. Ketoconazole blocks testosterone synthesis. Arch Intern Med 1982; 142:2137-40.
19. Pont A, Williams PL, Loose DS et al. Ketoconazole blocks adrenal steroid synthesis. Ann Intern Med 1982; 97:370-72.
20. Feldman D. Ketoconazole and other imidazole derivatives as inhibitors of steroidogenesis. Endocr Rev 1986; 7:409-20.
21. Korn ED, Von Brand T, Tobie EJ. The sterols of Trypanosoma cruzi and Crithidia fasciculata. Comp Biochem Physiol 1969; 30:601-10.
22. Docampo R. Biochemical and ultrastructural alterations produced by miconazole and econazole in Trypanosoma cruzi. Molec Biochem Parasitol 1981; 3:169-80.
23. Beach DH, Goad LJ, Holz Jr GG. Effects of ketoconazole on sterol biosynthesis by Trypanosoma cruzi epimastigotes. Biochem Biophys Res Commun 1986; 136:851-56.
24. Goad LJ, Berens RL, Marr JJ et al. The activity of ketoconazole and other azoles against Trypanosoma cruzi: Biochemistry and chemotherapeutic action in vitro. Molec Biochem Parasitol 1989; 32:179-90.
25. Urbina JA, Vivas J, Visbal G et al. Modification of the sterol composition of Trypanosoma (Schizotrypanum) cruzi epimastigotes by delta 24(25)-sterol methyl transferase inhibitors and their combinations with ketoconazole. Mol Biochem Parasitol 1995; 73:199-210.

26. Lazardi K, Urbina JA, de Souza W. Ultrastructural alterations induced by two ergosterol biosynthesis inhibitors, ketoconazole and terbinafine, on epimastigotes and amastigotes of Trypanosoma (Schizotrypanum) cruzi. Antimicrob Agents Chemother 1990; 34:2097-105.

27. Liendo A, Visbal G, Piras MM et al. Sterol composition and biosynthesis in Trypanosoma cruzi amastigotes. Mol Biochem Parasitol 1999; 104:81-91.

28. Liendo A, Lazardi K, Urbina JA. In-vitro antiproliferative effects and mechanism of action of the bis- triazole D0870 and its S(-) enantiomer against Trypanosoma cruzi. J Antimicrob Chemother 1998; 41:197-205.

29. Urbina JA, Lazardi K, Aguirre T et al. Antiproliferative synergism of the allylamine SF 86-327 and ketoconazole on epimastigotes and amastigotes of Trypanosoma (Schizotrypanum) cruzi. Antimicrob Agents Chemother 1988; 32:1237-42.

30. Urbina JA, Payares G, Contreras LM et al. Antiproliferative effects and mechanism of action of SCH 56592 against Trypanosoma (Schizotrypanum) cruzi: In vitro and in vivo studies. Antimicrob Agents Chemother 1998; 42:1771-77.

31. Buckner FS, Joubert BM, Boyle SM et al. Cloning and analysis of Trypanosoma cruzi lanosterol 14α-demethylase. Mol Biochem Parasitol 2003; 132:75-81.

32. Lepesheva GI, Zaitseva NG, Nes WD et al. CYP51 from Trypanosoma cruzi: A phyla-specific residue in the B' helix defines substrate preferences of sterol 14alpha-demethylase. J Biol Chem 2006; 281:3577-85.

33. Buckner F, Yokoyama K, Lockman J et al. A class of sterol 14-demethylase inhibitors as anti-Trypanosoma cruzi agents. Proc Natl Acad Sci USA 2003; 100:15149-53.

34. McCabe R. Failure of ketoconazole to cure chronic murine Chagas' disease. J Infect Dis 1988; 158:1408-09.

35. Urbina JA, Payares B, Molina J et al. Cure of short- and long-term experimental Chagas' disease using D0870. Science 1996; 273:969-71.

36. Moreira AA, de Souza HB, Amato-Neto V et al. Evaluation of the therapeutic activity of itraconazole in chronic infections, experimental and human, by Trypanosoma cruzi. Rev Inst Med Trop Sao Paulo 1992; 34:177-80.

37. Urbina JA, Payares G, Sanoja C et al. In vitro and in vivo activities of ravuconazole on Trypanosoma cruzi, the causative agent of Chagas disease. Int J Antimicrob Agents 2003; 21:27-38.

38. Urbina JA, Payares G, Sanoja C et al. Parasitological cure of acute and chronic experimental Chagas disease using the long-acting experimental triazole TAK-187. Activity against drug-resistant Trypanosoma cruzi strains. Int J Antimicrob Agents 2003; 21:39-48.

39. Corrales M, Cardozo R, Segura MA et al. Comparative efficacies of TAK-187, a long-lasting ergosterol biosynthesis inhibitor, and benznidazole in preventing cardiac damage in a murine model of Chagas' disease. Antimicrob Agents Chemother 2005; 49:1556-60.

40. Brener Z. An experimental and clinical assay with ketoconazole in the treatment of Chagas disease. Mem Inst Oswaldo Cruz 1993; 88:149-53.

41. Toledo MJ, Bahia MT, Veloso VM et al. Effects of specific treatment on parasitological and histopathological parameters in mice infected with different Trypanosoma cruzi clonal genotypes. J Antimicrob Chemother 2004; 53:1045-53.

42. Guedes PM, Urbina JA, de Lana M et al. Activity of the new triazole derivative albaconazole against Trypanosoma (Schizotrypanum) cruzi in dog hosts. Antimicrob Agents Chemother 2004; 48:4286-92.

43. Araujo MS, Martins-Filho OA, Pereira ME et al. A combination of benznidazole and ketoconazole enhances efficacy of chemotherapy of experimental Chagas' disease. J Antimicrob Chemother 2000; 45:819-24.

44. Maldonado RA, Molina J, Payares G et al. Experimental chemotherapy with combinations of ergosterol biosynthesis inhibitors in murine models of Chagas' disease. Antimicrob Agents Chemother 1993; 37:1353-59.

45. Benaim G, Sanders JM, Garcia-Marchan Y et al. Amiodarone has intrinsic anti-Trypanosoma cruzi activity and acts synergistically with posaconazole. J Med Chem 2006; 49:892-99.

46. Apt W, Aguilera X, Arribada A et al. Treatment of chronic Chagas' disease with itraconazole and allopurinol. Am J Trop Med Hyg 1998; 59:133-38.

47. Apt W, Arribada A, Zulantay I et al. Itraconazole or allopurinol in the treatment of chronic American trypanosomiasis: The results of clinical and parasitological examinations 11 years post-treatment. Ann Trop Med Parasitol 2005; 99:733-41.

48. Buckner FS, Verlinde CLMJ, La Flamme AC et al. Efficient technique for screening drugs for activity against Trypanosoma cruzi using parasites expressing β-galactosidase. Antimicrob Agents Chemother 1996; 40:2592-97.

49. McCabe RE, Remmington JS, Araujo FG. In vitro and in vivo effects of itraconazole against Trypanosoma cruzi. Am J Trop Med Hyg 1986; 35:280-84.
50. McCabe RE, Remington JS, Araujo FG. Ketoconazole inhibition of intracellular multiplication of Trypanosoma cruzi and protection of mice against lethal infection with the organism. J Infect Dis 1984; 150:594-601.
51. Urbina JA, Lazardi K, Marchan E et al. Mevinolin (lovastatin) potentiates the antiproliferative effects of ketoconazole and terbinafine against Trypanosoma (Schizotrypanum) cruzi: In vitro and in vivo studies. Antimicrob Agents Chemother 1993; 37:580-91.
52. Urbina JA, Lira R, Visbal G et al. In vitro antiproliferative effects and mechanism of action of the new triazole derivative UR-9825 against the protozoan parasite Trypanosoma (Schizotrypanum) cruzi. Antimicrob Agents-Chemother 2000; 44:2498-502.
53. Campos R, Amato NV, Moreira AA et al. Evaluation of the therapeutic activity of fluconazole in acute experimental infection caused by Trypanosoma cruzi. Rev Hosp Clin Fac Med Sao Paulo 1992; 47:174-75.
54. Molina J, Martins-Filho O, Brener Z et al. Activities of the triazole derivative SCH 56592 (posaconazole) against drug-resistant strains of the protozoan parasite Trypanosoma (Schizotrypanum) cruzi in immunocompetent and immunosuppressed murine hosts. Antimicrob Agents Chemother 2000; 44:150-55.

CHAPTER 7

Histone Deacetylases

David Horn*

Abstract

Deacetylation of histones is required for gene regulation and cell cycle progression and the mediators, the histone deacetylases, are being vigorously pursued as drug targets for cancer chemotherapy. The deacetylases are also potential drug targets against infectious diseases and genome sequencing revealed proteins of this class in each of three kinetoplastid parasites. These enzymes are now being characterised and assessed for chemotherapeutic potential.

Introduction

Many proteins rely upon post-translational modification for activity, subcellular localisation or to alter their structure and/or stabiltiy and some of these modifications are reversible. Reversible acetylation for example requires the combined action of acetylases (aka acetyltransferases) and deacetylases which work together to maintain the appropriate acetylation level. It was only just over ten years ago when these enzymes were first identified and in the case of the first deacetylase an inhibitor of the enzyme was used as a probe to purify the protein itself.[1] Reversible protein acetylation has been the subject of intense investigation ever since. Although research has focussed on histone (de)acetylation and its consequences, it is important to consider alternative in vivo substrates. Human 'histone' deacetylase 6 (HDAC6) is a tubulin deacetylase for example (see ref. 2).

A large number of putative histone acetylases and deacetylases have now been identified based on similarity to proteins with confirmed activity. Indeed, the completion of genome sequencing for protozoan parasites, including the three kinetoplastids, *Trypanosoma brucei*, *Trypanosoma cruzi* and *Leishmania major*, subsequently referred to as the Tritryps (see ref. 3), revealed that the conserved domains in these enzymes are pervasive among eukaryotes. Although studies on protein acetylation are at an early stage in kinetoplastids, work in other systems and current clinical studies suggest that putative (histone) deacetylases can be considered serious potential drug targets for the range of diseases caused by these parasites.

The Deacetylases

A number of histone modifications have been demonstrated but acetylation appears to be one of the most dynamic. Acetylation is most commonly on the ε amino groups of lysine residues near the histone *N*-termini within 'histone tails' that extend from the core of the nucleosome and beyond the DNA wrapped around the outer face of the histone octamer. The location of these modifications allows histone deacetylases to modulate chromatin structure and interaction[4] or remove binding platforms for regulatory factors (reviewed in ref. 5). The deacetylases themselves may be regulated by phosphorylation[6] and, in the case of the class II

*David Horn—London School of Hygiene & Tropical Medicine, Keppel Street, London, WC1E 7HT. UK. Email: david.horn@lshtm.ac.uk

Drug Targets in Kinetoplastid Parasites, edited by Hemanta K. Majumder.
©2008 Landes Bioscience and Springer Science+Business Media.

enzymes, shuttling between the cytoplasm and nucleus.[7] They are recruited to chromatin within multi-component complexes some of which contain additional catalytic components responsible for distinct histone modifications.[8] Histone acetylation can have targeted or more widespread effects when recruited to specific promoters or heterochromatin respectively for example and roles in cell cycle progression[9] and differentiation have been demonstrated.[7] In fact, all DNA-templated processes may depend upon appropriate histone modification patterns.

Five putative histone acetylases and seven deacetylases were revealed following Tritryp genome sequencing (see Table S4 in ref. 3). For comparison, *Saccharomyces cerevisiae* has ten deacetylases and humans have eighteen. These deacetylases are divided into three classes or super families (I-III) all of which are also found in bacteria. Classes I and II are structurally related and are distinct from the Sir2-related, NAD[+]-dependent class III enzymes (reviewed in ref. 10). The trypanosomatid deacetylases comprise two, two and three proteins from classes' I-III respectively. Small molecule modulators of class III enzyme activity have been identified (reviewed in ref. 11) but these compounds are not currently in clinical trials. I will therefore focus below on the four deacetylases from classes I and II and the inhibitors that target this class of enzymes, several of which are currently in clinical trials as anticancer agents. It is worth noting here that the acetylases may also emerge as targets for chemotherapy but small-molecule inhibitors of these enzymes have not yet been reported.

A phylogenetic analysis of the Tritryp Class I and II deacetylases (DACs; Fig. 1) reveals a number of interesting features. First, trypanosomatids branched very early from the eukaryotic

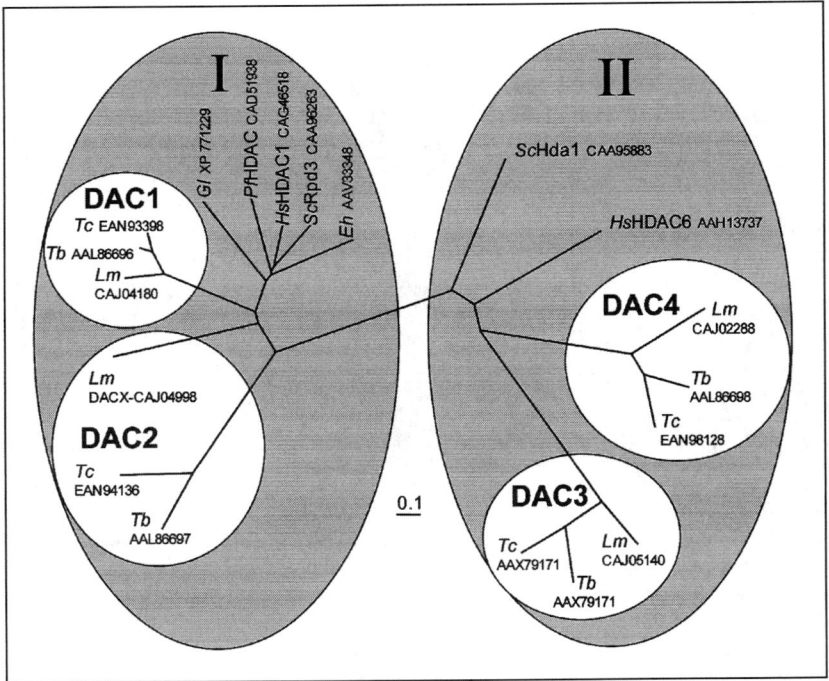

Figure 1. The phylogenetic relationship among the trypanosomatid deacetylases and other class I/II deacetylases. Related proteins were identified by BLAST analysis. The unrooted neighbour-joining tree was generated using Clustal 1.8X and TreeView. The two classes and the trypanosomatid enzymes (DAC1-4) are highlighted and all accession numbers are shown. *Tb, T. brucei; Tc, T. cruzi; Lm, L. major; Hs, Homo sapiens; Sc, S. cerevisiae; Gl, Giardia lamblia; Eh, Entamoeba hiostolytica; Pf, Plasmodium falciparum.*

lineage and the class I DACs clearly reflect this divergence. Second, the relationship between the *T. brucei, T. cruzi* and *Leishmania* proteins indicates the presence of putative DAC1, DAC3 and DAC4 homologues in all three parasites while *L. major* DAC2 does not cluster with trypanosome DAC2. This protein lacks several critical catalytic residues and was the only class I/II DAC in *T. brucei* that was dispensable and had no impact on cell cycle progression.[12]

The Inhibitors

The histone deacetylase inhibitors (HDi) are a structurally diverse group of naturally occurring and synthetic compounds (reviewed in ref. 13) that can induce growth arrest, differentiation and apoptosis of cancer cells ex vivo and in vivo. Some HDi are discussed below but for a more comprehensive description of the range of HDi see Table 4 in ref. 14. The first potent, small molecule HDi were described in 1995 (reviewed in ref. 15); trichostatin A from *Streptomyces* is a hydroxamic acid and a noncompetitive inhibitor of mammalian deacetylases that is effective in the nanomolar concentration range; trapoxin is a fungal cyclic tetra peptide and an irreversible inhibitor of mammalian deacetylases. Deacetylases were considered promising chemotherapy targets since the discovery of these first HDi and the subsequent demonstration of selectivity against several types of transformed cells.[14] Indeed, HDAC deregulation has been linked to cancer and several compounds have activity against a spectrum of tumours at doses that are well tolerated. Following positive results in a number of animal models, HDi are now one of the most promising classes of new anticancer agents and there are several different structural classes in clinical trials (reviewed in ref. 13). For example, the reversible inhibitor, suberoylanilide hydroxamic acid (SAHA) is in phase II clinical trials. HDi have also been considered potential chemotherapeutic agents against infectious protozoa (see ref. 16 and below) and have been extremely useful not only for identifying the first histone deacetylase (see above) but also for determining the consequences and importance of protein deacetylation in vivo.

Transformed and undifferentiated dividing cells are susceptible to HDi while nondividing cells appear to be resistant. Acetylated histones accumulate in normal nondividing cells but this appears to be tolerated. HDi may influence mammalian cell growth and division through a range of different pathways. Clearly, inhibition leads to the accumulation of inappropriately hyper-acetylated histones and other proteins thereby interfering with growth or cell cycle progression but the molecular consequences of deacetylase inhibition are not fully understood. HDi are generally thought to disrupt transcription control. For example histone hyper-acetylation may disrupt heterochromatin (reviewed in ref. 17) leading to up (or down)-regulation of a proportion of genes[18] with the associated secondary effects. The effects may be more pleiotropic however and histone hyper-acetylation may also compromise centromere function[19] or monitoring by the DNA repair and/or checkpoint machinery may lead to cell cycle arrest. Inappropriate acetylation could also impact upon the deposition of adjacent modifications[20] and the range of nonhistone targets may be significant. Indeed, HDi therapy may be more powerful than drugs that target a single molecular pathway because they affect several molecular programmes.

Intense interest in the potential applications of HDi has lead to the detailed characterisation of a range of compounds and the synthesis of a large number of derivatives. Among the known HDi and their derivatives, there are likely to be several compounds active against kinetoplastids. Not all deacetylases are equally sensitive to the different HDi so 'pan-specific' HDi may find utility against kinetoplastid parasites or kinetoplastid-isoform-specific compounds may emerge from the large chemical libraries now available.

Current Research

There is significant interest in using HDi against protozoan parasites. As described above, class I/II deacetylases have been identified in kinetoplastids but what about identifying HDi with activity against this group of organisms? The fungal metabolite, apicidin is a potent, reversible HDi active against the Apicomplexan parasites that cause malaria and toxoplasmosis.[16,21]

Although this particular HDi is inactive against kinetoplastids (see ref. 16) an apicidin analogue has been reported to have potent and selective activity against the African trypanosome, *T. brucei*.[22] In addition, one of the hydroxamate inhibitors displays a MIC$_{50}$ (minimum inhibitory concentration) of 1 μM in a four-day *T. brucei* growth assay (Charles Marson and DH, unpublished) and it seems likely that these compounds target DAC1, DAC3 or both (see below). It will now be important to further characterise the enzymes and to establish methods for assessing relative HDi activity.

As depicted in Figure 1, Tritryp genome sequencing revealed four class I/II deacetylases in each trypanosomatid. The genes encoding these enzymes (DAC1-4) have been analysed in *T. brucei*. This analysis revealed that two of the genes (DAC1 and DAC3) are essential for growth and another (DAC4) is required for normal cell cycle progression.[12] For comparison, both canonical class I (Rpd3) and class II HDACs (Hda1) are dispensable in *S. cerevisiae*.[23] Thus, DAC1 and DAC3 can be considered genetically validated as potential chemotherapy targets. These enzymes may also represent possible chemotherapy targets in other trypanosomatids (see Fig. 1). DAC4 may be considered a secondary target but cells lacking this gene display only a mild growth defect and a delay in the G$_2$/M phase of the cell cycle.[12] The structures of two HDi-bound deacetylases have been determined.[24,25] Using such structures as a template, molecular modelling of one of the essential proteins (*T. brucei* DAC1; Fig. 2) reveals the conserved substrate/inhibitor-binding pocket. Such models may facilitate the identification and/or design of more effective inhibitors.

The evolutionary divergence of trypanosomatids is reflected in molecular mechanisms that appear unusual compared to those characterised in humans and model organisms. Consistent with this trend, the transcription machinery and mechanism appear to be relatively simple (see ref. 3). Although histone acetylation is likely to be a pervasive feature of transcription

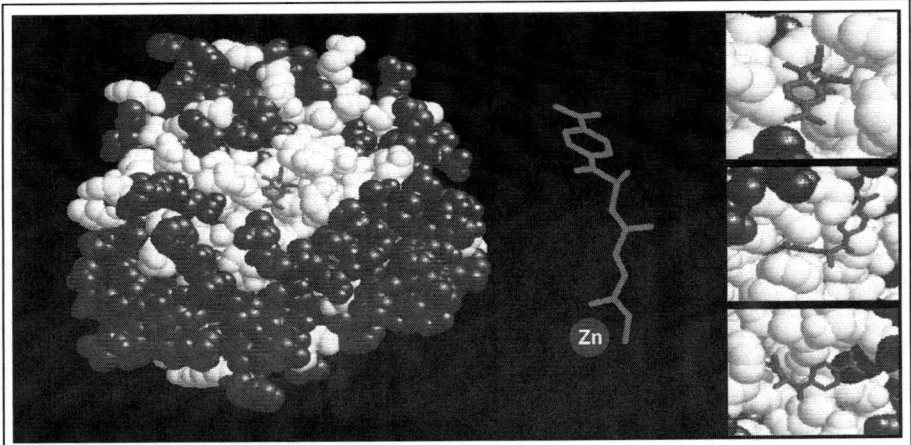

Figure 2. Molecular modelling of *Tb*DAC1. *Tb*DAC1 was modelled against the structure of an HDi-bound deacetylase (see the text) using Swiss-Model and the Swiss-Pdb Viewer[30] and the resulting space-filling model was viewed using RasMol.[31] Light residues are shared with human HDAC1 (see Fig. 1). Trichostatin A mimics the structure of acetyl-lysine and complexes with a zinc ion involved in acetamide cleavage at the active site (middle) and this HDi was manually docked into the substrate/inhibitor-binding pocket (left). Three close-up views of this region are shown on the right. This particular HDi may be inactive against DAC1 due to steric hindrance at the pocket entrance (see dark residue below the pocket in the image on the left) but it should be noted here that HDi and possibly natural substrates can alter protein structure in the vicinity of the pocket (see ref. 25).

regulation in eukaryotes, further analysis will be required to characterise DAC1-4 and to explore their role in transcription, replication, recombination and repair. The *T. brucei* DACs are currently being further characterised in my laboratory (Sam Alsford, Taemi Kawahara and DH, unpublished) and the lack of redundancy of DAC1 and DAC3 function facilitates this analysis. Conditional RNA interference is available to 'knock-down' specific mRNAs in *T. brucei* and not only confirms that these proteins are essential for growth but also provides material for the analysis of specific DAC function. These cells should allow roles in cell-cycle progression and in regulating gene expression to be defined. Regulated transcription has not been widely reported in kinetoplastids but we recently reported telomeric gene silencing in *T. brucei*.[26] It will be interesting to test the role of the class I/II DACs in this process especially since the canonical *S. cerevisiae* HDACs have a role in telomeric silencing[23] and HDi disrupt telomeric silencing in human cells.[27] Kinetoplastid histone *N*-terminal tails are sufficiently diverged relative to other eukaryotes such that most of the available histone modification-specific antisera will not cross-react (see ref. 28) but the strains described above may also allow the identification of specific (histone) substrates for each DAC once the appropriate antisera are available. In addition, determination of sub cellular localisation may indicate nuclear localisation consistent with histone substrates or cytoplasmic localisation possibly pointing to tubulin as a substrate for example (see ref. 2).

It will be particularly important to establish in vitro activity assays which can be used to assess candidate HDi. Commercial kits are available to facilitate such assays but it may be necessary to isolate certain DACs within native trypanosome complexes since cofactors may be required for activity (see ref. 29). It will then be possible to identify the enzymes targeted by a particular HDi and then to correlate that activity with *T. brucei* growth inhibition. HDi with increased potency may be identified using this approach and compounds active against *T. brucei* for example will be good candidates for testing against other kinetoplastids. It may even be possible to identify effective compounds already registered for use in humans.

Summary

HDi are a relatively new class of therapeutic agent but they currently show great promise as effective cancer therapies. Some existing compounds show activity against parasitic protozoa including kinetoplastids and additional, more potent and/or specific compounds may emerge from new or existing libraries. Molecular modelling of the essential kinetoplastid proteins (see Fig. 2) should facilitate the selection or design and synthesis of improved compounds and HDi may subsequently be developed as effective therapies for infectious diseases. In addition to therapeutic potential, small-molecule HDi will be useful chemical probes for further characterisation of the kinetoplastid DACs. Experimental analysis has only just begun for kinetoplastids however and it will be important to fully characterise the DACs and assess candidate compounds using a combination of pharmacological and biochemical assays.

Acknowledgements

I would like to thank David Warhurst (LSHTM) for help with molecular modelling and Charles Marson (University College London), Shane Wilkinson, Taemi Kawahara and Sam Alsford (LSHTM) for helpful comments on the manuscript. Work in my laboratory is supported by The Wellcome Trust (069909).

References

1. Taunton J, Hassig CA, Schreiber SL. A mammalian histone deacetylase related to the yeast transcriptional regulator Rpd3p. Science 1996; 272:408-11.
2. Hubbert C, Guardiola A, Shao R et al. HDAC6 is a microtubule-associated deacetylase. Nature 2002; 417:455-8.
3. Ivens AC, Peacock CS, Worthey EA et al. The genome of the kinetoplastid parasite, Leishmania major. Science 2005; 309:436-42.

4. Shogren-Knaak M, Ishii H, Sun JM et al. Histone H4-K16 acetylation controls chromatin structure and protein interactions. Science 2006; 311:844-7.
5. Loyola A, Almouzni G. Bromodomains in living cells participate in deciphering the histone code. Trends Cell Biol 2004; 14:279-81.
6. Pflum M, Tong J, Lane W et al. Histone deacetylase I phosphorylation promotes enzymatic activity and complex formation. J Biol Chem 2001; 276:47733-41.
7. McKinsey TA, Zhang CL, Lu J et al. Signal-dependent nuclear export of a histone deacetylase regulates muscle differentiation. Nature 2000; 408:106-11.
8. Yang L, Mei Q, Zielinska-Kwiatkowska A et al. An ERG (ets-related gene)-associated histone methyltransferase interacts with histone deacetylases 1/2 and transcription corepressors mSin3A/B. Biochem J 2003; 369:651-7.
9. Pile L, Schlag E, Wassarman D. The SIN3/RPD3 deacetylase complex is essential for G_2 phase cell cycle progression and regulation of SMRTER corepressor levels. Mol Cell Biol 2002; 22:4965-76.
10. Blander G, Guarente L. The sir2 family of protein deacetylases. Annu Rev Biochem 2004; 73:417-35.
11. Grubisha O, Smith BC, Denu JM. Small molecule regulation of Sir2 protein deacetylases. Febs J 2005; 272:4607-16.
12. Ingram AK, Horn D. Histone deacetylases in Trypanosoma brucei: Two are essential and another is required for normal cell cycle progression. Mol Microbiol 2002; 45:89-97.
13. Dokmanovic M, Marks PA. Prospects: Histone deacetylase inhibitors. J Cell Biochem 2005; 96:293-304.
14. Johnstone RW. Histone-deacetylase inhibitors: Novel drugs for the treatment of cancer. Nat Rev Drug Discov 2002; 1:287-99.
15. Yoshida M, Horinouchi S, Beppu T. Trichostatin A and trapoxin: Novel chemical probes for the role of histone acetylation in chromatin structure and function. Bioessays 1995; 17:423-30.
16. Darkin-Rattray SJ, Gurnett AM, Myers RW et al. Apicidin: A novel antiprotozoal agent that inhibits parasite histone deacetylase. Proc Natl Acad Sci USA 1996; 93:13143-7.
17. Taddei A, Roche D, Bickmore WA et al. The effects of histone deacetylase inhibitors on heterochromatin: Implications for anticancer therapy? EMBO Rep 2005; 6:520-4.
18. Van Lint C, Emiliani S, Verdin E. The expression of a small fraction of cellular genes is changed in response to histone hyperacetylation. Gene Expr 1996; 5:245-53.
19. Ekwall K, Olsson T, Turner BM et al. Transient inhibition of histone deacetylation alters the structural and functional imprint at fission yeast centromeres. Cell 1997; 91:1021-32.
20. Eissenberg JC, Elgin SC. Antagonizing the neighbours. Nature 2005; 438:1090-1.
21. Andrews KT, Walduck A, Kelso MJ et al. Anti-malarial effect of histone deacetylation inhibitors and mammalian tumour cytodifferentiating agents. Int J Parasitol 2000; 30:761-8.
22. Murray PJ, Kranz M, Ladlow M et al. The synthesis of cyclic tetrapeptoid analogues of the antiprotozoal natural product apicidin. Bioorg Med Chem Lett 2001; 11:773-6.
23. Rundlett SE, Carmen AA, Kobayashi R et al. HDA1 and RPD3 are members of distinct yeast histone deacetylase complexes that regulate silencing and transcription. Proc Natl Acad Sci USA 1996; 93:14503-8.
24. Finnin MS, Donigian JR, Cohen A et al. Structures of a histone deacetylase homologue bound to the TSA and SAHA inhibitors. Nature 1999; 401:188-93.
25. Somoza JR, Skene RJ, Katz BA et al. Structural snapshots of human HDAC8 provide insights into the class I histone deacetylases. Structure 2004; 12:1325-34.
26. Glover L, Horn D. Repression of polymerase I-mediated gene expression at Trypanosoma brucei telomeres. EMBO Rep 2006; 7:93-9.
27. Baur J, Zou Y, Shay J et al. Telomere position effect in human cells. Science 2001; 292:2075-7.
28. Horn, D. Nuclear gene transcription and chromatin in Trypanosoma brucei. Int J Parasitol 2001; 31:1157-65.
29. Wu J, Carmen AA, Kobayashi R et al. HDA2 and HDA3 are related proteins that interact with and are essential for the activity of the yeast histone deacetylase HDA1. Proc Natl Acad Sci USA 2001; 98:4391-6.
30. Guex N, Peitsch MC. SWISS-MODEL and the Swiss-PdbViewer: An environment for comparative protein modeling. Electrophoresis 1997; 18:2714-23.
31. Sayle RA, Milner White EJ. RASMOL: Biomolecular graphics for all. Trends Biochem Sci 1995; 20:374-6.

CHAPTER 8

Targeting Glycoproteins or Glycolipids and Their Metabolic Pathways for Antiparasite Therapy

Sumi Mukhopadhyay nee Bandyopadhyay and Chitra Mandal*

Abstract

Carbohydrate-based therapy, known as glycotherapeutics, is a new and emerging field that promises to be the future hope for combating kinetoplastid infections more efficiently and effectively. Targeting novel glycoproteins/lipids, which are important disease determinants of kinetoplastid diseases, have helped in the development of this field. Better and refined understanding of all the available data would possibly help us in providing a future direction for rational drug design and better disease management. This review intends to focus on such lines, which will give us an insight into the future hope for development of novel therapeutic strategies through glycobiological platform for combating kinetoplastid infections.

An Introduction to the Kinetoplastid Parasites

Today's Kinetoplastida form a diverse order of flagellated protozoans that have evolved from an ancient lineage, rooted near the base of the eukaryotic tree. The disease caused by Kinetoplastida has always plagued mankind, and today most are as prevalent as they have ever been. Kinetoplastid parasites cause disease in humans, animals and plants, severely affecting human health. Sleeping sickness (caused by pathogenic subspecies of African *Trypanosoma brucei*), Chagas disease or South American trypanosomiasis (caused by *Trypanosoma cruzi*) and the Leishmaniases (caused by *Leishmania sp*) are the major human diseases caused by kinetoplastids. In addition to their medical importance, kinetoplastid parasites also cost developing nations millions of dollars in lost agricultural revenues, since other kinetoplastids are pestilences that strike agricultural produce from crops, to fish to cattle.[1]

Trypanosomiasis

Trypanosomiasis is a vector-borne parasitic disease. *Trypanosoma*, the parasites concerned, are protozoa transmitted to humans by tsetse flies (*glossina*). There have been three severe epidemics in Africa over the last century. At present the disease has reappeared in endemic form in several foci. According to the World Health Organization (WHO), 60 million people in 36 countries of sub-Saharan Africa are infected. Sleeping sickness has become the first or second greatest cause of mortality, ahead of HIV/AIDS.

*Corresponding Author: Chitra Mandal—Immunobiology Division, Indian Institute of Chemical Biology, 4 Raja S.C. Mullick Road, Jadavpur, Kolkata -700 032, India. Email: cmandal@iicb.res.in or chitra_mandal@yahoo.com

Drug Targets in Kinetoplastid Parasites, edited by Hemanta K. Majumder.
©2008 Landes Bioscience and Springer Science+Business Media.

The complex life cycle of *T. brucei* involves the alternation between the vertebrate host and the tsetse fly vector. In the tsetse fly, the parasite resides in the procyclic form or as epimastigote, here it differentiates into the metatrypansome form, the form that can infect the vertebrate host. In the host they survive as trypomastigote, extracellularly in the blood, once again picked up by fly and begin the cycle anew.

T. cruzi, the trypomastigote form, may remain extracellularly in the blood or may reside intracellular as amastigote. The rest of the cycle is being same as *T. brucei.* Without treatment, the disease is fatal.

Four drugs are currently in use. Suramine, discovered in 1921, is used in the initial phase of treatment of *T.b. rhodesiense.* Pentamidine, discovered in 1941, is used in treatment of the initial phase of *T.b. gambiense* sleeping sickness.

Second phase treatments includes Melarsoprol and Eflornithine, discovered in 1949 and 1990, are available to treat the advanced stage of sleeping sickness, no matter which parasite is the cause. It is the last arsenical derivative in existence with drug-resistance in 30% patients of central Africa.

Leishmaniasis

Leishmaniasis, a vector borne disease caused by obligate intracellular macrophage protozoa, is characterized by complexity and diversity. The disease can present itself in four different forms in humans: cutaneous, diffuse cutaneous, mucocutaneous, and visceral.

Worldwide, there are 2 million new cases including children (<5 years) each year and 1/10 of the world's population is at risk of infection.[2,3] Mortality rate increased from 41,000 to 59,000 deaths from 2000 to 2001.[4] The true incidence and prevalence of leishmaniasis is uncertain because many cases go undiagnosed or unreported in the area where the infection is endemic.

The disease is endemic to about 88 countries.[5] Up to half of the world's cases of VL (Kala azar) occur in India and over 90% live in the Indian state of Bihar.[6] Over the past decade major epidemics have been reported from eastern India and Bangladesh, Sudan, and northeastern Brazil.

Coinfection with *Leishmania* and human immunodeficiency virus (HIV) is emerging as a new and frightful disease and is becoming increasingly frequent.[7] WHO estimates that one-third of HIV infected persons (~17 million) live in the zones of endemic leishmaniasis infection.[8]

The disease is mainly transmitted by phlebotomine sandflies. Parasitized female sandfly takes a blood meal from a human host and thus promastigote forms enter in host. They are ingested by host macrophage and metamorphose into amastigote forms and reproduce by binary fission. They increase in number until the cell eventually bursts, then infect other phagocyctic cells and continue the cycle.

Treatment of leishmanisis includes the classical drug, sodium stibolgluconate that hinder with the carbohydrate energy metabolism of the parasite. Diamidines (pentamidine) inhibit the trypanothione metabolism by interacting with the DNA minicircle, while Allopurinol inhibits the incorporation of the adenoisiner into DNA. Paromomycin inhibit the protein synthetic machinery and Amphotericin B reacts with ergosterol in the membrane, increases the permeability of plasma membranes to ions and small molecules. Miltefosine, a phosphocholine analogue, is showing promising results by affecting cell signaling pathway and membrane synthesis.[9] Unfortunately, all available therapeutic arsenals are injected parenterally. Antimonials are no longer effective in Northeast India. Traditional second-line drugs (pentamidine and amphotericin B) are limited by toxicity and availability; and newer formulations of amphotericin B are effective but simply unaffordable in developing countries. Miltefosine appears effective in patients with high level of antimonial resistance. Due to fear of drug resistance, the possible use of combination therapies needs to be explored.[10]

Considering the wide global prevalence of the kinetoplastid infections and the growing resistance towards the present drugs in use, it is becoming important to discover newer alternative drugs. An attempt towards this direction has been the carbohydrate-based drugs. Following sections provide a glimpse into this journey.

Why Target Glycoproteins and Glycolipids?

Glycocojugates (proteins/lipids) present on the protozoal surface have been demonstrated to play an important role in determining parasite survival and infectivity.[11-13] These glycocojugates can thus possibly serve as candidates for drug targets (Table 1). Glycocojugates mediate adhesion of pathogen-host cell, therefore, synthetic analogues designed to mimic such

Table 1. Occurrence of glycoconjugates in kinetoplastid parasites

Species	Possible Target	Properties	References
Trypanosoma brucei	Variant surface glycoprotein (VSG)	Coat glycoprotein of bloodstream trypomastigote	16-18
	Procyclic acidic repetitive protein/Procyclins (PARP)	Coat glycoprotein of procyclic forms	30
Trypanosoma cruzi	Mucins	Dense and continuous coat composed of a layer of type I glycosylinositolphospholipids	21-25
	1. 35-50 kDa	Epimastigote and metacyclic trypomastigote	
	2. 80-200 kDa	Cell culture-derived trypomastigotes	
	3. Trans sialidase	α2,3-linked sialic acid residues are transferred from host glycoconjugates	
	LPPG	Cell surface glycan of epimastigote	31
	Glycosylinositolphospholipids	Cell surface glycolipids	21-22
Leishmania sp	Lipophosphoglycan (LPG)	Coat glycoprotein present in promastigote but downregulated expression in amasigotes..	39
	Glycosylinositolphospholipids (GIPLs)	Protein free glycolipids. Major constituents of the amastigote surface	61-63
	gp63	Major cell surface glycoprotein. Expressed in lower level in amastigotes	43
	secreted acid phosphatase (sAP)	glycosylated proteins Secreted from the flagella in promastigotes.	64-65
	Proteophosphoglycans (PPG)	GPI-anchored present in promastigotes and amastigotes.	66-68
	Phosphoglycan	Hydrophilic phosphoglycan consisting of capped oligosaccharide repeat units but minus the GPI anchor and the glycan core	69
	Sialoglycans	Sialic acid derivatives present both on promastigotes and amastigotes, but Neu5Gc present only on the amastigote	48-50

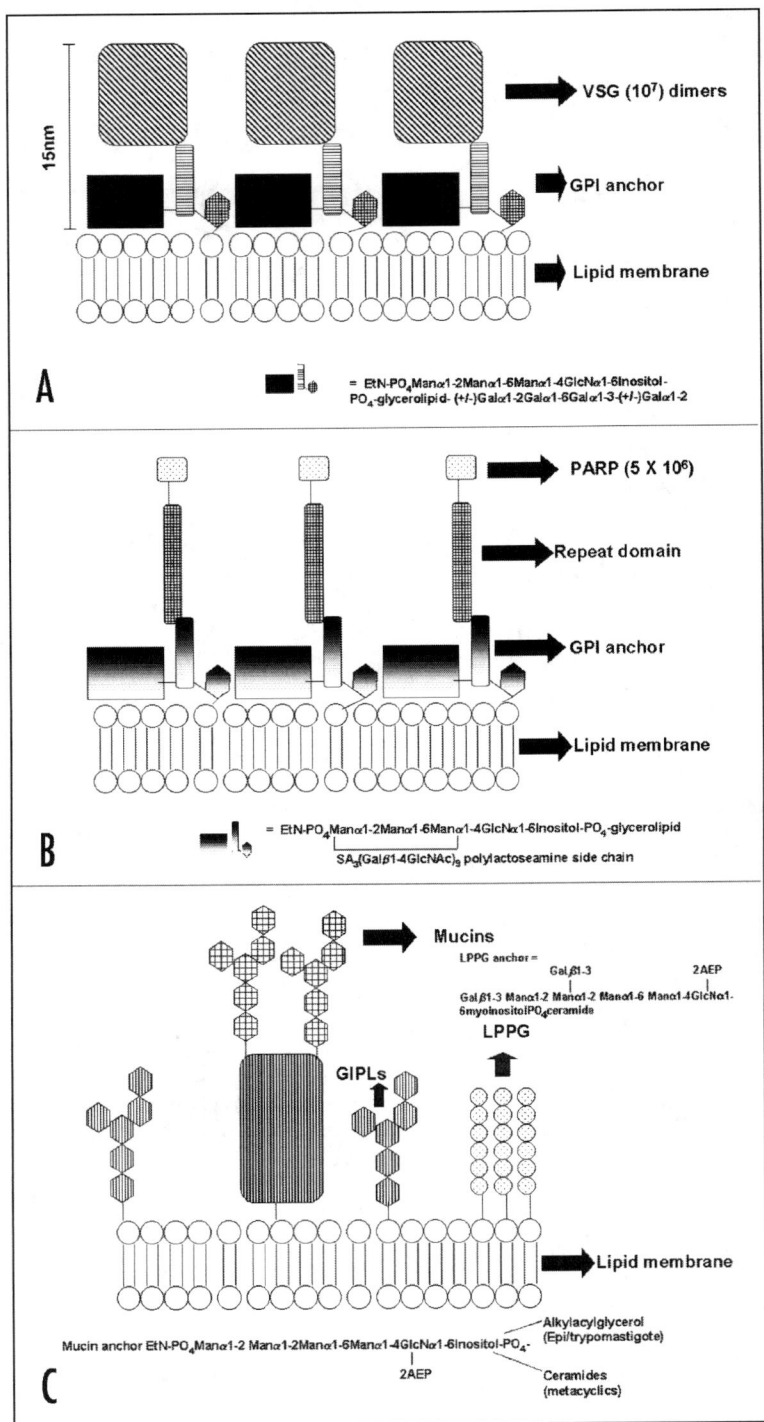

Figure 1. Please see legend on following page.

Figure 1, viewed on previous page. Schematic representation of the major surface glycoconjugates of *Trypanosoma brucei*. A) Metacyclic. Each molecule of variant surface glycoprotein (VSG) consists of two GPI-anchored *N*-glycosylated monomers. B) Procyclic. The surface is densely covered with procyclic acidic repetitive proteins (PARPs). These are GPI-anchored polypeptides with polyanionic repeat domains. The repeat domain could be either glutamate-proline repeats or glycine-proline- glutamate-glutamate-threonine repeat sequences. The surface of PARP is attached with $Man_5GlcNAc_2$ oligosaccharide. C) *Trypanosoma cruzi*. The cell surface of *T. cruzi* is covered with a dense layer of glycosylinositolphospholipids (GIPLs), mucins, and lipopeptidophosphoglycan (LPPG). Aminoethylphosphonate is AEP.

glycocojugates can be used to prevent the adhesion and subsequent infection. Expression of glycocojugates is mainly due to up-regulation of certain enzymes (Table 1), hence designing novel enzyme-inhibitors to block the metabolic pathway also appears a novel proposition.

To fight against drug-induced pathogen-resistance, synthetic analogues offer a major advantage, as these analogs are similar to their natural ligands, pathogens are less likely to develop resistance against these analogues and may be an alternative novel therapeutic agent.

In light of the prominent role played by glycocojugates present on the kinetoplastid protozoa, it becomes essential to study their occurrence, metabolism and biological roles. This might help us in unraveling their role in disease-biology and further in designing effective drugs. A brief discussion about the disease-biology, current treatment protocols, disease-status, the major glycoconjugates and glycoconjugate-based drug development on two major kinetoplastid infections *viz*, leishmaniasis and trypanosomes are discussed below.

Glycoconjugates of *Trypanosomes*

Most of the major cell-surface proteins of these organisms are anchored to the plasma membrane via glycosyl-phosphatidylinositol (GPI) anchors, though a few of the plasma membrane proteins use trans-membrane polypeptide anchors (Fig. 1).

The cell surface of the metacyclic *T. brucei* trypomastigote is covered with a dense glycocalyx composed of approximately ten million variant surface glycoprotein (VSG, Fig. 1A). The surface of the procyclic form is densely covered with procyclic acidic repetitive proteins (PARPs). These are also GPI-anchored polypeptides with polyanionic repeat domains (Fig. 1B). The cell surface of *T. cruzi* is covered with a dense layer of glycosylinositolphospholipids (GIPLs), mucins, and lipopeptidophosphoglycan (LPPG) (Fig. 1C).

Biological Relevance of Glycoconjugates in Trypanosomiasis

The VSG glycocalyx provides a diffusion barrier, thus preventing complement-mediated lysis. The various PARPs serve as lectin-binding ligands within the fly. The surface coat primarily seems to have a protective function and the sialylation of the mucins provide extra ability to survive in different environments. Sialylation is proposed to reduce the susceptibility of the parasite to anti-α-Gal antibodies present in the mammalian bloodstream.[14] The heavily sialylated coat may also provide a structural barrier to other lytic agents encountered by the parasite as well as promote adherence to the macrophage and modulate the production of NO and cytokines.[15]

Taking into consideration the pivotal role played by these glycoconjugates, substantial work on the development of anti-trypanosomal drugs have been reported. A brief account of these investigations are described in the following section.

Targeting VSG Epitopes

About 1000 VSG genes are expressed sequentially, allowing the parasite to evade recognition by the humoral immune system. This strong ability to mutate creates difficulty not only for the host's immune system but also for the development of drug therapies. Each monomer of VSG protein carries at least one occupied N-linked glycosylation site.[16-18] The mature anchor consists of the classical GPI core ethanolamine-HPO$_4$-6Manα1,2)Man(α1,6)Man(α1,4)GlcN(α1,6)-inositoldimyristoylglycerol.

VSG expose only highly variable and immuno-dominant epitopes to the immune system, whereas conserved epitopes become inaccessible for large molecules. Reducing the size of binders, less immunogenic, cryptic VSG epitopes forms an obvious solution to combat these parasites. This goal was achieved by introducing dromedary heavy-chain antibodies with high specificity for the conserved Asn-linked carbohydrate of VSG, recognizing epitopes common to multiple VSG classes by penetrating the dense coat to target their epitope. The employment of this binder as a molecular recognition unit in immuno-toxins designed for trypanosomosis therapy was demonstrated as specific trypanolysis.[19]

Towards the development of analogues targeted towards VSG, myristate analogs have been reported. Unlike mammalian GPIs, the diacylglycerol moiety of the VSG anchor contains only myristate (tetradecanoate), making it a unique drug target. Myristate analogs are useful for studying the mechanism of GPI myristolyation, and they are also candidates for anti-trypanosomal chemotherapy.[20]

Targeting Mucin Glycoproteins

T. cruzi has a dense and continuous coat composed of a layer of type I GIPLs,[21-22] and a family of small mucins[23-25] projecting above the GIPL layer (Fig. 1C). These mucins are heavily glycosylated with O-linked GlcNAc as the internal residue and further modified with one to five galactosyl residues. Mucin-like glycoproteins, 35-50 kDa and 80-200 kDa are expressed on the parasitic forms found in the insect and cell culture-derived trypomastigotes respectively.[26] Glycoproteins like gp90 have been exploited for development of drugs.[27]

The expression of gp90, a stage-specific surface glycoprotein trypomastigotes, is inversely correlated with the parasite's ability to invade mammalian cells. Anti-sense oligonucleotides complementary to a region of the gp90 gene selectively inhibit gp90 synthesis in the metacyclic forms, which inhibited their entry into target cells. Parasites were incubated with a phosphorothioate oligonucleotide based on a sequence of the gp90 coding strand (PS1) or PS2, the anti-sense counterpart of PS1; containing phosphodiester linkages. PS2 but not PS1 inhibited the expression of gp90. The gp90 mRNA levels were diminished in PS2-treated parasites. Treatment with PS2 did not affect the expression of other surface glycoproteins involved in parasite-host cell interaction. Expression of gp90 was also inhibited by other phosphorothioate oligonucleotides targeted to the gp90 gene.[27]

Lactose Derivatives Are Inhibitors of Trans-Sialidase Present in the Mucins

Sialic acids typically present as terminal residues on glycoproteins and glycolipids are known to play a significant role in the mediation of many biological phenomena involving cell-cell and cell-matrix interactions by either reacting with specific surface receptors or masking other carbohydrate recognition sites.[28]

Trypanosomes are unable to synthesize sialic acid de novo and have the unusual ability to acquire it from host glycoconjugates. The galactose-rich mucin is the acceptor for the trans-sialylation in which α2,3-linked sialic acid residues are transferred from host-glycoconjugates by a trans-sialidase (TcTS) enzyme to the terminal β-Gal residues of the mucin. Surface sialic acid helps in protection of the parasite against the lysis by complement and in cell invasion. Therefore, enzyme inhibitors to trans-sialidase can significantly reduce the parasite burden and can thus effectively control the infection. More recently, Lactitol, the lactose open chain derivatives, have been reported to be good inhibitors of trans-sialidase activity. Lactitol inhibit resialylation of parasite mucins when incubated with live trypanosomes and diminish the infection in cultured cells by 20-27% indicating that compounds directed to the lactose-binding site might be good inhibitors of TcTS.[29]

Molecular cloning and understanding the function of genes regulating trans-sialidases would therefore have profound influence in developing new chemotherapeutic approaches to combat these protozoan infections. As we continue to further our understanding of sialoglycans in parasitic protozoa, it will help to foster innovative new strategies for diminishing the mortality and morbidity caused by these parasites.

PARPs

These are largely acidic glycoproteins.[30] Each cell expresses approximately 5 million copies of procyclins (Fig. 1B). The PARPs form a dense glycocalyx of GPI anchors with the polyanionic polypeptide repeat domains projecting above the membrane.

LPPG

LPPG is the major cell surface glycan of epimastigote with approximately 1.5×10^7 copies per cell.[31] LPPG can be considered a member of the GPI family based on the presence of the Man(α1,4)GlcN(α1,6)*myo*-inositol-1-PO$_4$-lipid motif, the hallmark of all GPI anchors.

Glycoconjugate Biosynthetic Machinery in Trypanosome

Continuous effort to explore glycoconjugates in trypanosomes to develop drugs is ongoing. In this context, the biosynthetic pathway of these glycoconjugates also appears a lucrative area for designing anti-trypanosome drugs.[32] Utilization of GPIs for anchoring proteins to the plasma membrane is essential for the trypanosomes. Specific inhibitors of parasite-GPI biosynthetic enzymes could thus be designed for use as therapeutic agents. A schematic representation of the biosynthetic machinery[32] operating in trypanosomes for the production of these essential glyconjugates is provided in Figure 2.

Targeting Enzymes of the Biosynthetic Pathway

The natural substrate for α-D-mannosyl transferase of GPI biosynthesis is D-GlcNα1-6-D-myo-inositol-1-P-sn-1-2-diacylglycerol. A diastereoisomer, D-GlcNα1-6-L-myo-inositol-1-P-sn-1-2-diacylglycerol, is an inhibitor of this enzyme in a trypanosomal cell-free system. The L-myo-inositol-1-phosphate is the principal inhibitory component.[33] Comparisons between the natural substrate and the inhibitors suggested that the inhibitors bind to the first α-D-mannosyltransferase. Unfortunately, none of the L-myo-inositol-containing compounds that inhibited GPI biosynthesis in a parasite cell-free system had any effect on biosynthesis in human cell-free system, suggesting that other related parasite-specific inhibitors of this essential pathway need to be developed.

Further studies in this direction are the development of GPI anchor biosynthesis inhibitors like mannosamine.[34] Mannosamine exerts these effects by becoming incorporated into GPI anchor intermediates. Trypanosomes were biosynthetically labeled with [3H]mannosamine, the main glycolipid metabolite of mannosamine was shown to be ManN-Man-GlcN-PI. A trypanosome cell-free system preloaded with this compound was significantly impaired in its ability to synthesize GPI anchor intermediates beyond Manα1-6Manα1-4GlcNα1-6PI. This compound is, therefore, proposed to be an inhibitor of the Dol-P-Man: Manα1-6Manα1-4GlcNaα1-6PIα1-2-mannosyltransferase of the GPI biosynthetic pathway. In living trypanosomes, mannosamine (4 mM) able to reduce the rate of formation of mature GPI anchor precursors by 80%.

Glycoconjugates of *Leishmania*

Leishmania like all parasite, survive and proliferate in highly hostile environments and have evolved special mechanisms that enable them to endure these adverse conditions. To protect themselves from such harsh conditions one of the adaptive mechanism includes the production of a dense cell surface glycocalyx composed of lipophosphoglycan (LPG, composed of phosphatidyl(*myo*)inositol lipid anchor, glycan core, Gal(β1,4)Man(α1)-PO$_4$ backbone repeat

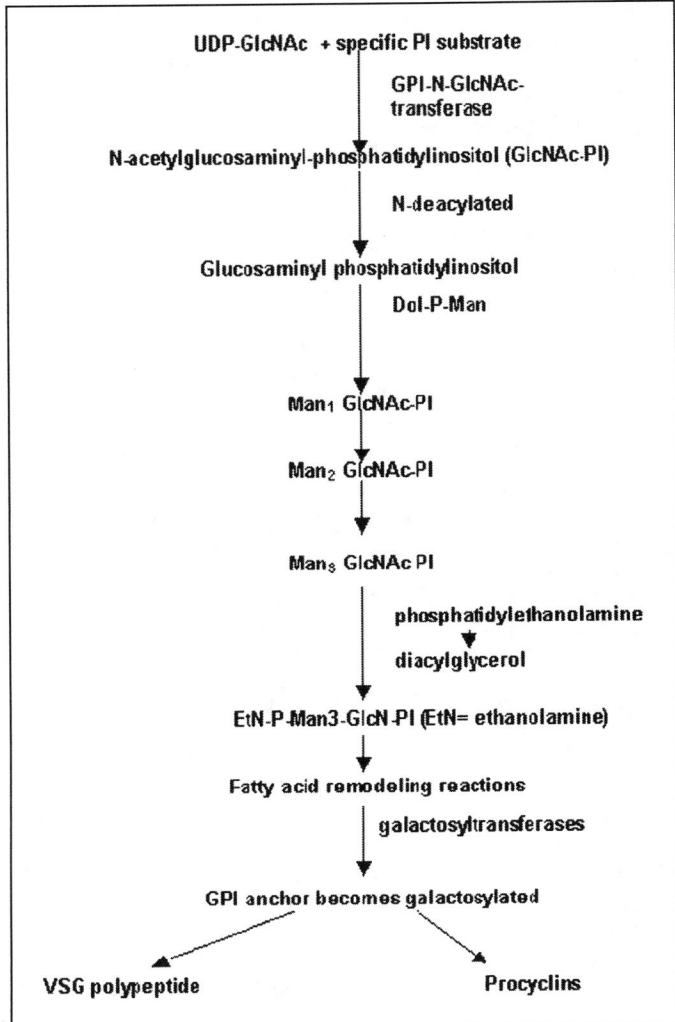

Figure 2. Schematic representation of GPI biosynthetic pathway in *Trypanosoma sp.* After biosynthesis of GPI anchor it is galactosylated and the VSG/procyclins polypeptide is subsequently added.

units oligosaccharide cap structure), glycosylinositolphospholipids, or GIPLs (Protein free glycolipids)[35-36] and secreted glycoconjugates, proteophosphoglycan (PPG)[37-38] and secreted acid phosphatase (sAP)[39-42] (Fig. 3A,B).

Biological Relevance

LPG prevents complement-mediated lysis of the promastigote and serves as a ligand for receptor-mediated endocytosis by the macrophage. LPG of amastigotes inhibits protein kinase C and the microbicidal oxidative burst as well as phagosome-endosome fusion. GIPLs have a role in macrophage invasion and are involved in modulating signaling events in the

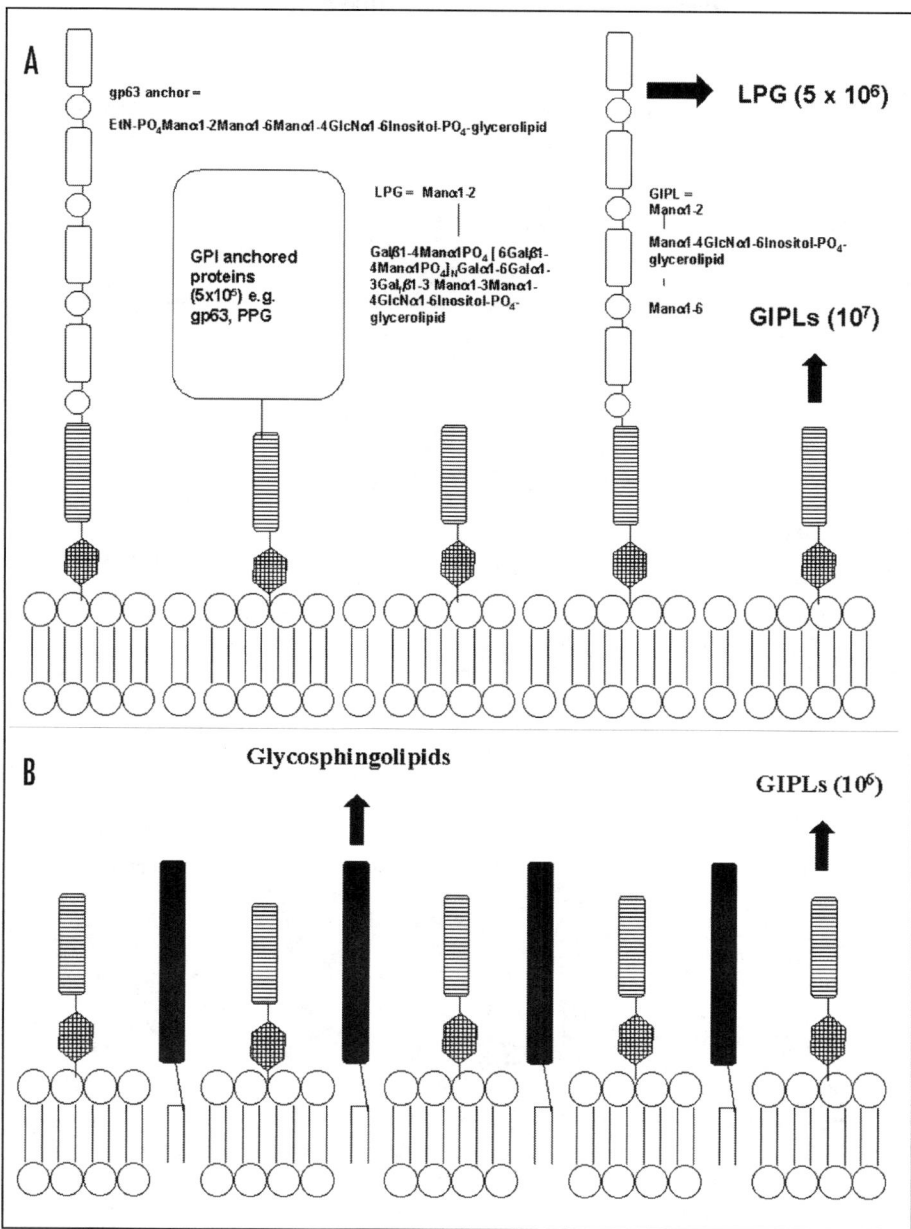

Figure 3. Schematic representation of the cell surfaces of *Leishmania sp.* A) Promastigote. Only the major cell surface molecules are depicted approximate copy numbers per cell are indicated in parentheses. The cell surface contains several abundant GPI anchored proteins (gp63, PPG), the polydisperse lipophosphoglycans (LPG) and the low-molecular-mass glycoinositolphospholipids (GIPLs). B) Amastigote. The LPG and the major surface glycoprotein is greatly down-regulated and the plasma membrane contains a number of glycosphingolipids.[36]

macrophage. gp63 serves as a ligand for the macrophage receptor protect against complement mediated lysis. PPG forms gel-like matrix, and these interlocking filaments, traps the parasites in the sand fly anterior gut. In the amastigotes, they activate the complement system via the mannose-binding pathway. Therefore, targeting a few of these glycoconjugates appears to be a novel proposition. A brief introduction of the reported attempts of discovering novel glycoconjgate based anti-leishmanials is outlined below.

Development of Anti-Leishmanials by Targeting

gp63

It is a 63-kDa zinc metalloprotease and is anchored to the cell surface via a myristic acid containing GPI anchor (Fig. 3A). This major cell surface glycoprotein on promastigotes with 500,000 copies per cell is accounting for 1% of all cellular proteins. In amastigotes, it is expressed to a lower level and the bulk is found in the flagellar pocket, as opposed to covering the entire surface on promastigotes.[43] The crystal structure revealed to contain an active site structural motif found in other zinc proteases that may aid design of specific inhibitors.[44] gp63 contains three potential glycosylation sites and the glycans are biantennary high mannose-type, and some bear a terminal Glc in α1,3 linkage. The $Man_6GlcNAc_2$ and $GlcMan_6GlcNAc_2$ are two major structures on all promastigote species. In amastigotes, the structures are more variable. Whether the stage-specific changes in glycan structure affect parasite infectivity and development is unknown.

A fibronectin-like molecule, gp63, plays a key role in parasite-macrophage interaction. Binding of gp63 to macrophage receptors is inhibited by Arg-Gly-Asp-Ser-containing synthetic peptides of fibronectin and by antibodies to these peptides.[45]

Recent reports have also unraveled novel peptide inhibitors of leishmania gp63 based on the cleavage site of myristoylated alanine-rich C kinase substrate (MARCKS)-related protein.[46] However, efficient non-toxic inhibitors of gp63 do not exist. MARCKS and MARCKS-related protein (MRP) are protein kinase C substrates in various cells, including macrophages and an excellent substrate for gp63. A major cleavage site was identified within the MRP effector domain (ED) and the synthetic ED peptide (MRP-ED) inhibit MRP hydrolysis and may be a good gp63 inhibitors. As phosphorylation of ED serine residues prevent gp63-mediated MRP degradation, a pseudophosphorylated peptide was synthesized in which serine residues were substituted by aspartate (3DMRP-ED). This 3DMRP-ED is a highly effective inhibitor of both soluble and parasite-associated gp63. Finally, MRP-ED peptides were synthesized together with an N-terminal HIV-1 Tat transduction domain (TD) to obtain cell-permeant peptide constructs. Such peptides retain gp63 inhibitory activity and efficiently enter both macrophages and parasites in a Tat TD-dependent manner. These studies provide the basis for developing potent cell-permeant inhibitors of gp63.[46]

gp63 Producing Enzymes

Peptidomimetics, at low micromolar concentrations are able to inhibit in vitro cleavage of a synthetic peptide substrate by purified gp63 from *L. major*. Development of higher affinity metalloprotease inhibitors may provide a novel avenue for treatment of parasitic diseases.[47]

Galactofuranose on LPG

LPGs is found in all species of *Leishmania* and localized over the entire promastigotes surface including the flagellum. It is composed of four domains: 1-*O*-alkyl-2-*lyso*-phosphatidyl(*myo*)inositol lipid anchor, glycan core, Gal(β1,4)Man(α1)-PO_4 backbone repeat units and an oligosaccharide cap structure.[38,41] Structural analysis of LPG from different species has revealed complete conservation of the lipid anchor, the glycan core, and the Gal (β1, 4) Man (α1)-PO_4 backbone of repeat units. Galactofuranose has been characterized in the middle of the oligosaccharide core linked β1-3 to Man. The presence in protozoan parasites of galactose in the furanose configuration is a feature, which deserves further attention since the mammalian hosts do not appear to produce glycoconjugates containing this structural unit.

The metabolic pathways involved in the attachment to or removal of galactofuranose from glycoconjugates is an area of incipient research, but of growing importance, since it will foster the design of inhibitors, which may prove to be useful for the treatment of disease.[35]

Sialoglycans

The topography of *Leishmania* parasites with regard to their sialoglycan profile remains a poorly investigated area and it is only recently that the status of sialoglycans on *Leishmania donovani* promastigotes as well as amastigotes has been reported from the authors laboratory.[48-50] Sialic acids are a structurally complex family of nine-carbon polyhydroxy amino ketoacid of *N*-(Neu5Ac) and *O*-substituted derivatives of neuraminic acid (Neu5,9Ac$_2$).[51] They have been demonstrated to have diverse roles.[52-58] The promastigotes and amastigotes contain 7×10^5 and 12.8×10^5 molecules of Neu5Ac per cell demonstrating a 2.0 fold higher copy number in amastigotes.[48-49] Although the presence of Neu5Ac and Neu5Gc are detectable on amastigotes, Neu5Gc is absent on promastigotes. Neu5Gc is a major derivative in most mammals, including our closest evolutionary relatives; the great apes and thought to be absent in healthy humans thereby make it a potential candidate for new drug design.[49]

The examination of surface density of sialoglycoconjugates present on *L. donovani* indicates the predominance of α2-6 linked sialylglycotopes on the promastigote, though α2-3 also present. In contrast, α2-3 linked sialylglycotopes is higher on amastigotes. It remains to be investigated whether these sialoglycans are playing a major role in the infectivity and intracellular survival of the parasite.

The presence of α2-6 linked sialoglycoproteins on promastigotes is quite different than 164 and 150 kDa on amastigote. While five α2-3 linked sialoglycans (130, 117, 106, 70, and 61 kDa) are identified, amastigotes adsorb different sets of sialoglycans (188, 162, 136, 137 and 124 kDa). This raises the possibility that during transformation, parasites acquire a new array of sialoglycans onto their surface.[48-49] Additionally, two *O*-acetylated (Neu5,9Ac$_2$) sialoglycoproteins (123 and 109 kDa) on promastigotes and 164 and 150 kDa on the amastigote cell surface have also been demonstrated.

These sialoglycans induce high humoral responses in the host and these antibodies have important application in the field of therapy as well they have anti-leishmanial activity as reported by its capacity to induce complement lysis of parasite.[60] Thus, sialoglycans appear to be important target molecules.

Other Glycoconjugates

There are four more major glycoconjugates (GIPLs, *sAP*, PPG, Phosphoglycan) reported on the cell surface of *leishmania* sp. Considering their important biological roles, they might also serve as important drug targets. Accordingly, a brief introduction is outlined below:

GIPLs

The GIPLs are a major family of low molecular weight glycolipids synthesized by parasites, which are not attached to proteins or polysaccharides.[61-63] These are expressed in very high copy numbers, approximately 10^7 copies per cell on both promastigote and amastigote surfaces (Fig. 3).

sAP

With the exception of *L. major*, all *Leishmania* promastigotes secrete sAP from the flagellar pocket[64-65] and tend to form distinct macromolecular complexes with proteoglycans. The sAP peptides are heavily glycosylated and they are phosphodiester-linked to serine residues and commonly consist of the 6Gal(β1,4)Man(α1-)PO$_4$ repeats units found on LPG. The average number of repeat units is 32.

PPG

Proteophosphoglycans in promastigotes include filamentous PPG or fPPG and a putative GPI-anchored cell-associated form or mPPG.[66-68] Amastigotes secrete their own nonfilamentous and stage-specific form termed aPPG. Compositionally fPPG consists of 95% phosphoglycans, with an abundance of serine, alanine, and proline in the peptide component. Over 80-90% of the serine residues are phosphoglycosylated with short Gal-Man-PO$_4$ repeats attached via phosphodiester bonds, which are terminated by small oligosaccharide cap structures.

Phosphoglycan

Culture supernatants of *Leishmania* promastigotes contain a hydrophilic phosphoglycan consisting of capped oligosaccharide repeat units identical to those found on LPG, but minus the GPI anchor and the glycan core.[69]

Leishmania internalization takes place through receptors present on the host macrophage; parasite glycoconjugates serve as important ligands for their entry into the host. Two of the glycoconjugates purified from promastigotes are potent inhibitors of internalization, 75% inhibition being obtained using fucose-mannose glycoproteic ligand (6-10 µg/ml) and phosphate mannogalactan ligand. The simultaneous presence of both ligands in low concentrations yielded an increase in inhibitory activity above that found for each ligand alone, indicating that promastigotes may use at least two receptor sites for penetration into macrophages.[70]

Glycosylphosphatidylinositol (GPI) Lipids

Glycosylphosphatidylinositol (GPI) structures are attached to many cell surface glycoproteins in lower and higher eukaryotes. GPI structures are particularly abundant in trypanosomatid parasites where they are attached to complex phosphosaccharides, glycoproteins and mature surface glycolipids. The high density of GPI structures at all life-cycle stages Leishmania suggests that the GPI-biosynthetic pathway might be a reasonable target for the development of antiparasite drugs. It has been reported that that synthetic analogues of early GPI intermediates having the 2-hydroxyl group of the D-myo-inositol residue methylated can get incorporated in the subsequent steps of the GPI biosynthetic pathways of *leishmania major* but not in the human (HeLa) cells. These findings suggest that the discovery and development of specific inhibitors of parasite GPI biosynthesis are attainable goals.[71]

Enzymes Producing Glycoproteins

Recent studies towards the development of anti-leishmania drugs has unraveled the critical enzyme GDP-mannose pyrophosphorylase as a drug target in *L. mexicana*.[72] Leishmania synthesize a range of mannose-containing glycoconjugates thought to be essential for virulence in the mammalian host and sandfly vector. A prerequisite for the synthesis of these molecules is the availability of the activated mannose donor, GDP-Man, the product of the catalysis of mannose-1-phosphate and GTP by GDP-mannose pyrophosphorylase (GDP-MP). In contrast to the lethal phenotype in fungi, the deletion of the gene in *L. mexicana* does not affect parasite viability but lead to a total loss of virulence, making GDP-MP an ideal target for anti-leishmania drug development.

Among the most relevant and well-studied parasite glycotopes are the sialic acid group of sugars and their derivatives. Expression of surface terminal sialic acid and its derivative is due to the interplay between several enzymes namely sialidases, sialyltransferases, O-acetyl transferase and O-acetyl esterase. Therefore, synthetic analogues design to inhibit their expression appears to be an attractive treatment protocol.

As Drug Carrier

Glycoconjugates have also been reported to serve as carriers for receptor-mediated drug targeting in the treatment of experimental visceral leishmaniasis (VL). For example Methotrexate (MTX) coupled to mannosyl bovine serum albumin (BSA) has been taken up efficiently through the mannosyl receptors present on macrophages. Binding experiments indicate that conjugation does not decrease the affinity of the neoglycoprotein for its cell surface receptor. The drug conjugate eliminate intracellular amastigotes of *L.donovani* in mouse peritoneal macrophages about 100 times more efficiently than free drug on the basis of 50% inhibitory dose. Inhibitory effect of the conjugate is directly proportional to the density of sugar on the neoglycoprotein carrier. Colchicine and monensin, inhibitors of receptor-mediated endocytosis, can prevent the leishmanicidal effect of the conjugate. Anti-leishmanial effect of the conjugate can be competitively inhibited by mannose-BSA and mannan. In experimental model, the drug conjugate reduces the spleen parasite burden by more than 85% whereas the same concentration of free drug cause little effect. These results indicate that MTX-neoglycoprotein conjugate binds specifically to macrophages, and are internalized and degraded in lysosomes releasing the active drug to act on leishmania parasites.[73]

Perspective

In light with the growing interest in the field of carbohydrates and its various applications in therapy, identification, characterization of newly induced carbohydrate epitopes on the parasite appears to be an important area of investigation. This can help in unraveling important biomarkers, which can subsequently be exploited in the development of new reliable, cheap, easy and rapid assays. Further these molecules may also be important tools in the development of carbohydrate-based therapy in the field of infectious diseases.

Acknowledgements

The Department of Biotechnology and the Indian Council of Medical Research, Govt. of India supported the work. Dr. S. Mukhopadhyay nee Bandyopadhyay worked as a Senior Research Fellow of Council of Scientific and Industrial Research. We acknowledge Dr. A. Toreno (Instituto de salud Carlos III Majadahonda, Madrid, Spain) and Dr. R. Vlasak (University of Salzburg, Austria) for their gifts of reagents. Our special thanks are to Dr. GJ. Gerwig and Prof JP Kamerling (Bijvoet Center for Biomolecular Research, Utrecht University, Netherlands) for their help with the mass spectrometric and HPLC studies. Our sincere thanks are also for Dr. PR. Crocker (Wellcome Trust Biocentre, Dundee,UK), Prof.Dr.R. Schauer (Biohemishes Institut, Kiel, Germany), Prof. R. Brossmer and Dr. R. Schwartz-Albiez (University of Heidelberg, Germany), for helping us with different sialic acid probes. Dr. S. Hinderlich (Institut für Molekularbiologie und Biochemie Freie Universität Berlin, Germany) is acknowledged for enzyme assay. Dr. M. Chatterjee (Institute of PG Med Edn & Res, Kolkata 700 020) and Dr. AK. Chava are acknowledged for their valuable contribution towards this work and finally, we also thank Mr. A. Mullick, for his excellent technical assistance.

References

1. Davila AM, Tyler KM. Combating Kinetoplastid diseases. Kinetoplastid Biol Dis 2002; 1:6.
2. Handman E. Leishmaniasis: Current status of vaccine development. Clin Microbiol Rev 2001; 14:229-243.
3. Bhattacharya SK, Jha TK, Sundar S et al. Efficacy and tolerability of miltefosine for childhood visceral leishmaniasis in india. Clin Infect Dis 2004; 38:217-221.
4. Guerin PJ, Olliaro P, Sundar S et al. Visceral leishmaniasis: Current status of control, diagnosis, and treatment, and a proposed research and development agenda. Lancet Infect Dis 2002; 2:494-501.
5. Davies CR, Kaye P, Croft SL et al. Leishmaniasis: New approaches to disease control. BMJ 2003; 326:377-382.
6. Sundar S, Agarwal G, Ria M et al. Treatment of Indian visceral leishmaniasis with single dose or daily infusion of low dose liposomal amphotericin B, a randomized trial. BMJ 2002; 323:419-422.
7. Iqbal J, Hira PR, Saroj G et al. Imported visceral leishmaniasis: Diagnostic dilemmas and comparative analysis of three assays. J Clin Microbiol 2002; 40:475-479.
8. Alvar J, Avate CC, Rrez-solar BG et al. Leishmania and human immunodeficiency virus coinfection: The first 10 years. Clin Microbiol Rev 1997; 10:298-319.
9. Harder A, Greif G, Haberkorn A. Chemotherapeutic approaches to protozoa: Kinetoplastida-current level of knowledge and outlook. Parasitol Res 2001; 87:778-780.

10. Rosenthal E, Marty P. Recent understanding in the treatment of visceral leishmaniasis. JPGM 2003; 49:61-68.

11. Chava AK, Bandyopadhyay S, Chatterjee M et al. Sialoglycans in protozoal diseases: Their detection,modes of acquisition and emerging biological roles. Glycoconj J 2004; 20:199-206.

12. Sinha D, Chatterjee M, Mandal C. O-acetylation of sialic acids- their detection, biological significance and alteration in diseases- a Review: Trends Glycosci Glycotechnol 2000; 12:17-33.

13. Guha-Niyogi A, Sullivan DR, Turco SJ. Glycoconjugate structures of parasitic protozoa. Glycobiology 2001; 11:45-59.

14. Pereira-Chioccola VL, Acosta-Serrano A, de Almeida IC et al. Mucin-like molecules form a negatively charged coat that protects Trypanosoma cruzi trypomastigotes from killing by human anti-α-galactosyl antibodies. J Cell Sci 2000; 113:1299-1307.

15. de Diego J, Punzon C, Duarte M et al. Alteration of macrophage function by a Trypanosoma cruzi membrane mucin. J Immunol 1997; 159:4983-4989.

16. Zamze SE, Ashford DA, Wooten EW et al. Structural characterization of the asparagine-linked oligosaccharides from Trypanosoma brucei type II and type III variant surface glycoproteins. J Biol Chem 1991; 266:20244-20261.

17. Mehlert A, Zitzman N, Richardson JM et al. The glycosylation of variant surface glycoproteins and procyclic acidic repetitive proteins of Trypanosoma brucei. Mol Biochem Parasitol 1998; 91:145-152.

18. Ferguson MAJ. The structure, biosynthesis and functions of glycosylphosphatidylinositol anchors, and the contributions of trypanosome research. J Cell Sci 1999; 112:2799-2809.

19. Stijlemans B, Conrath K, Cortez-Retamozo V et al. Efficient targeting of conserved cryptic epitopes of infectious agents by single domain antibodies: African trypanosomes as paradigm. J Biol Chem 2004; 279:1256-1261.

20. Doering TL, Lu T, Werbovetz KA et al. Toxicity of myristic acid analogs toward African trypanosomes. Proc Natl Acad Sci 1994; 91:9735-9739.

21. Lederkremer RM, Lima C, Ramirez MI et al. Complete structure of the glycan of lipopeptidophosphoglycan from Trypanosoma cruzi epimastigotes. J Biol Chem 1991; 266:23670-23675.

22. Carriera JC, Jones C, Wait R et al. Structural variations in the glcosylinositolphospholipids of different strains of Trypanosoma cruzi. Glycoconj J 1996; 13:955-966.

23. Schenkman S, Ferguson MAJ, Heise N et al. Mucin-like glycoproteins linked to the membrane by glycosylphosphatidylinositol anchors are the major acceptors of sialic acid in a reaction catalyzed by trans-sialidase in metacyclic forms of Trypanosoma cruzi. Mol Biochem Parasitol 1993; 59:293-304.

24. Previato JO, Jones C, Xavier MT et al. Structural characterization of the major glycosyl-phosphatidylinositol membrane-anchored glycoprotein from epimastigote forms of Trypanosoma cruzi Y-strain. J Biol Chem 1995; 270:7241-7250.

25. Serrano AA, Schenkman S, Yoshida N et al. The lipid structure of the glycosylphosphatidylinositol-anchored mucin-like sialic acid acceptors of Trypanosoma cruzi changes during parasite differentiation from epimastigotes to infective metacyclic trypomastigote forms. J Biol Chem 1995; 270:27244-27253.

26. Almeida IC, Ferguson MAJ, Schenkman S et al. Lytic anti α-galactosyl antibodies from patients with chronic Chagas' disease recognize novel O-linked oligosaccharides on mucin-like glycosylphosphatidylinositol anchored glycoproteins of Trypanosoma cruzi. Biochem J 1994; 304:793-802.

27. Malaga S, Yoshida N. Targeted reduction in expression of Trypanosoma cruzi surface glycoprotein gp90 increases parasite infectivity. Infect Immun 2001; 69:353-359.

28. Kelm S, Schauer R. Sialic acids in molecular and cellular interactions. Int Rev Cytol 1997; 175:137-240.

29. Agusti R, Paris G, Ratier L et al. Lactose derivatives are inhibitors of Trypanosoma cruzi trans-sialidase activity toward conventional substrates in vitro and in vivo. Glycobiology 2004; 14:659-670.

30. Treumann A, Zitzman N, Husmeier A et al. Structural characterization of two forms of procyclic acidic repetitive protein expressed by procyclic forms of Trypanosoma brucei. J Mol Biol 1997; 269:529-547.

31. de Lederkremer RM, Colli W. Galactofuranose-containing glycoconjugates in trypanosomatids. Glycobiology 1995; 5:547-552.

32. Mcconville MJ, Ferguson MAJ. The structure, biosynthesis and function of glycosylated phosphatidylinositols in the parasitic protozoa and higher eukaryotes Biochem J 1993; 294:305-324.

33. Smith TK, Sharma DK, Crossman A et al. Parasite and mammalian GPI biosynthetic pathways can be distinguished using synthetic substrate analogues. EMBO J 1997; 16:6667-6675.
34. Ralton JE, Milne KG, Guther ML et al. The mechanism of inhibition of glycosylphosphatidylinositol anchor biosynthesis in Trypanosoma brucei by mannosamine. J Biol Chem 1993; 268:24183-24189.
35. Turco SJ, Descoteaux A. The lipophosphoglycan of Leishmania parasites. Annu Rev Microbiol 1992; 46:65-94.
36. McConville MJ, Blackwell JM. Developmental changes in the glycosylated phosphatidylinositols of Leishmania donovani: Characterization of the promastigote and amastigote glycolipids. J Biol Chem 1991; 266:15170-15179.
37. Ilg T, Overath P, Ferguson MA et al. O- and N-glycosylation of the Leishmania mexicana-secreted acid phosphatase: Characterization of a new class of phosphoserine-linked glycans. J Biol Chem 1994; 269:24073-24081.
38. Ilg T, Stierhof YD, Wiese M et al. Characterization of phosphoglycan-containing secretory products of Leishmania. Parasitology 1994; 108(Suppl):S63-71.
39. Lovelace JK, Gottlieb M. Comparison of extracellular acid phosphatases from various isolates of Leishmania. Am J Trop Med Hyg 1986; 35:1121-1128.
40. Ilg T, Montgomery J, Stierhof YD et al. Molecular cloning and characterization of a novel repeat-containing Leishmania major gene, ppg1, that encodes a membrane-associated form of proteophosphoglycan with a putative glycosylphosphatidylinositol anchor. J Biol Chem 1999; 274:31410-31420.
41. Ilg T, Handman E, Stierhof YD. Proteophosphoglycans from Leishmania promastigotes and amastigotes. Biochem Soc Trans 1999; 4:518-525.
42. Mukhopadhyay nee Bandyopadhyay S, Mandal C. Glycobiology of leishmania donovani. Indian J Med Res 2006; 123:203-220.
43. Medina-Acosta E, Kavess RE, Schwartz H et al. The promastigote surface protease (gp63) of Leishmania is expressed but differentially processed and localized in the amastigote stage. Mol Biochem Parasitol 1989; 37:263-273.
44. Schlagenhauf E, Etges R, Metcalf P. The crystal structure of the Leishmania major surface proteinase leishmanolysin (gp63). Structure 1998; 6:1035-1046.
45. Soteriadou KP, Remoundos MS, Katsikas MC et al. The Ser-Arg-Tyr-Asp region of the major surface glycoprotein of Leishmania mimics the Arg-Gly-Asp-Ser cell attachment region of fibronectin. J Biol Chem 1992; 267:13980-13985.
46. Corradin S, Ransijn A, Corradin G et al. Novel peptide inhibitors of Leishmania gp63 based on the cleavage site of MARCKS (myristoylated alanine-rich C kinase substrate)-related protein. Biochem J 2002; 367:761-769.
47. Bangs JD, Ransom DA, Nimick M et al. In vitro cytocidal effects on Trypanosoma brucei and inhibition of Leishmania major GP63 by peptidomimetic metalloprotease inhibitors. Mol Biochem Parasitol 2001; 114:111-117.
48. Chatterjee M, Chava AK, Kohla G et al. Identification and characterization of adsorbed serum sialoglycans on Leishmania donovani promastigotes. Glycobiology 2003; 5:351-361.
49. Chava AK, Chatterjee M, Gerwig GJ et al. Identification of sialic acids on Leishmania donovani amastigotes. Biol Chem 2004; 385:59-66.
50. Chava AK, Chatterjee M, Mandal C. O-acetyl sialic acids in parasitic diseases. In: Yarema KJ, ed. Chapter 3 in Hand book of Carbohydrate Engineering. USA: Published by Taylor and Francis Group, book division, 2005:71-97.
51. Schauer R. Achievements and challenges of sialic acid research. Glycoconj J 2000; 17:485-499.
52. Chava AK, Chatterjee M, Sundar S et al. O-acetyl sialioglycoconjugates on erythrocytes for diagnosis and prognosis of Indian Visceral leishmaniasis and its biological role. Trends and Research in leishmaniasis, 2005; 5:223-243.
53. Chava AK, Chatterjee M, Sharma V et al. Differential expression of O-acetylated sialoglycoconjugates induces a variable degree of complement-mediated hemolysis in Indian leishmaniasis. J Infect Dis 2004; 189:1257-1264.
54. Bandyopadhyay S, Chatterjee M, Sundar S et al. Identification of 9-O-acetylated sialoglycans on peripheral blood mononuclear cells in Indian visceral leishmaniasis. Glycoconj J 2004; 20:531-536.
55. Chava AK, Chatterjee M, Sundar S et al. Development of an assay for quantification of linkage-specific O-acetylated sialoglycans on erythrocytes; its application in Indian visceral leishmnaiasis. J Immunol Meth 2002; 270:1-10.
56. Chatterjee M, Sharma V, Mandal C et al. Identification of antibodies directed against O-Acetylated sialic acids in Visceral Leishmaniasis: Its diagnostic and prognostic role. Glycoconj J 1998; 15:1141-1147.

57. Chatterjee M, Basu K, Basu D et al. Distribution of IgG subclasses in antimonial unresponsive Indian kala-azar patients. Clin Exp Immunol 1998; 114:408-413.
58. Sharma V, Chatterjee M, Mandal C et al. Rapid diagnosis of visceral leishmaniasis using Achatinin-H, a 9-O-acetylated sialic acid binding lectin. Amer J Trop Med Hyg 1998; 58:551-554.
59. Bandyopadhyay S, Chatterjee M, Pal S et al. Purification, characterization of O-acetylated sialoglycoconjugatesspecific IgM, and development of an enzyme-linked immunosorbent assay for diagnosis and follow-up of Indian visceral leishmaniasis patients. Diagn Microbiol Infect Dis 2004; 50:15-24.
60. Bandyopadhyay S, Chatterjee M, Das T et al. Antibodies directed against O-acetylated sialoglycoconjugates accelerate complement activation in leishmania donovani promastigotes. J Infect Dis 2004; 190:2010-2019.
61. McConville MJ, Ferguson MA. The structure, biosynthesis and function of glycosylated phosphatidylinositols in the parasitic protozoa and higher eukaryotes. Biochem J 1993; 294:305-324.
62. Turco SJ. Glycoproteins of parasites. In: Montreul J, Vliegenhart JFG, Schachter H, eds. Glycoproteins and Disease. Elsevier Science, B.V., 1996:113-124.
63. Ferguson MA. The structure, biosynthesis and functions of glycosylphosphatidylinositol anchors, and the contributions of trypanosome research. J Cell Sci 1999; 112:2799-2809.
64. Bates PA, Hermes I, Dwyer DM. Golgi mediated post translational processing of secretory acid phosphatase by Leishmania donovani promastigotes. Mol Biochem Parasitol 1990; 39:247-256.
65. Stierhof YD, Ilg T, Russell DG et al. Characterization of polymer release from the flagellar pocket of Leishmania mexicana promastigotes. J Cell Biol 1994; 125:321-331.
66. Ilg T, Overath P, Ferguson MA et al. O- and N-glycosylation of the Leishmania mexicana-secreted acid phosphatase: Characterization of a new class of phosphoserine-linked glycans. J Biol Chem 1994; 269:24073-24081.
67. Ilg T, Stierhof YD, Wiese M et al. Characterization of phosphoglycan-containing secretory products of Leishmania. Parasitology 1994; 108:S63-71.
68. Ilg T, Handman E, Stierhof YD. Proteophosphoglycans from Leishmania promastigotes and amastigotes. Biochem Soc Trans 1999; 4:518-25.
69. Turco SJ, Descoteaux A. The lipophosphoglycan of Leishmania parasites. Annu Rev Microbiol 1992; 46:65-94.
70. Palatnik CB, Borojevic R, Previato JO et al. Inhibition of Leishmania donovani promastigote internalization into murine macrophages by chemically defined parasite glycoconjugate ligands. Infect Immun 1989; 57:754-763.
71. Smith TK, Sharma DK, Crossman A et al. Parasite and mammalian GPI biosynthetic pathways can be distinguished using synthetic substrate analogues. EMBO J 1997; 16:6667-6675.
72. Davis AJ, Perugini MA, Smith BJ et al. Properties of GDP-mannose pyrophosphorylase, a critical enzyme and drug target in Leishmania mexicana. J Biol Chem 2004; 279:12462-12468.
73. Chakraborty P, Bhaduri AN, Das PK. Neoglycoproteins as carriers for receptor-mediated drug targeting in the treatment of experimental visceral leishmaniasis. J Protozool 1990; 37:358-364.

CHAPTER 9

DNA Topoisomerases of *Leishmania*:
The Potential Targets for Anti-Leishmanial Therapy

Benu Brata Das, Agneyo Ganguly and Hemanta K. Majumder*

Abstract

Protozoan parasites of the genus *Leishmania* cause severe diseases that threaten human beings, both for the high mortality rates involved and the economic loss resulting from morbidity, primarily in the tropical and subtropical areas. This ancient eukaryote shows variable genetic diversity in their life cycle, wherein DNA topoisomerases play a key role in cellular processes affecting the topology and organization of intracellular DNA. Kinetoplastid topoisomerases offer most attractive targets for their structural diversity from other eukaryotic counterparts and their indispensable function in cell biology. Therefore, understanding the biology of kinetoplastid topoisomerases and the components and steps involved in this intricate process provide opportunities for target based drug designing against protozoan parasitic diseases.

Introduction

Leishmaniasis is a disease complex caused by 17 different species of protozoan parasites belonging to the genus *Leishmania*. The parasites are transmitted between mammalian hosts by phlebotomine sandflies. There are an estimated 12 million humans infected, with an incidence of 0.5 million cases of the visceral forms of the disease and 1.5 to 2.0 million cases of the cutaneous form of the disease. Leishmaniasis has a worldwide distribution with important foci of infection in Central and South America, Southern Europe, North and East Africa, the Middle East, and the Indian subcontinent. Currently the main foci of visceral leishmaniasis (VL) are in Sudan and India and those of cutaneous leishmaniasis (CL) are in Afghanistan, Syria, and Brazil. In addition to the two major clinical forms of the diseases, VL and CL, there are other cutaneous manifestations, including mucocutaneous leishmaniasis (MCL), diffuse cutaneous leishmaniasis (DCL), recidivans leishmaniasis (LR), and post-kala-azar dermal leishmaniasis (PKDL) that are often linked to host immune status. The number of cases of leishmaniasis is probably under estimated as leishmaniasis is a reportable disease in only 40 of the 88 countries where it is known to be present.[1] Although the global burden of leishmaniasis has remained stable for several years, the patterns of the disease change continiously. With increasing numbers of human immunodeficiency virus (HIV) coinfections, human migration, and resettlement, there is a possibility of resurgence of the disease.[1] Improved approaches to diagnosis, vaccine development, vector and reservoir control and new drugs for treatment are still required.

To make the situation even worse, some parasite strains have also developed resistance against the classical antimonial drugs, like sodium stibogluconate and megalumine antimonite. The

*Corresponding Author: Hemanta K. Majumder—Molecular Parasitology Laboratory, Indian Institute of Chemical Biology, 4 Raja S.C Mullick Road, Kolkata-700032, India. Email: hkmajumder@iicb.res.in

Drug Targets in Kinetoplastid Parasites, edited by Hemanta K. Majumder.

second line of drugs, amphotericin B and pentamidines, although used clinically are very toxic. Therefore improved chemotherapy of leishmanial infection is still desirable and the need for new molecular targets on which to base the future treatment strategies is clearly justified. In search for such strategies DNA topoisomerases of *Leishmania* offer most attractive targets. The aim of this article is to provide an insight into the target based therapeutic approach against Leishmaniasis.

DNA Topoisomerases: The Wonder Enzyme

Topoisomerases are enzymes that use DNA strand scission, manipulation, and rejoining activities to directly modulate DNA topology. These actions provide a powerful means to effect changes in DNA supercoiling levels and allow some topoisomerases to both unknot and decatenate chromosomes. They are truly wonders, as in their presence, DNA strands can pass each other as if the physical boundaries between them have disappeared. They single handedly solve various topological problems for effective propagation of the genetic material. They are involved in replication, transcription, chromosomal condensation and segregation and many other vital cellular processes.[2] The immense interest in topoisomerase research in recent years derives not only from the recognition of their crucial role in managing DNA topology, but also from one major advance in the field. A wide variety of topoisomerase-targeted drugs have been identified, many of which generate cytotoxic lesions by trapping the enzymes in covalent complexes on the DNA. These topoisomerase poisons include both anti-microbials, antiparasitic and anti-cancer chemotherapeutics, some of which are currently in widespread clinical use.

Classification of DNA Topoisomerases

Topoisomerases are divided into two classes based primarily on their mode of cleaving DNA.[3] Type I DNA topoisomerases act by making a transient nick on a single strand of duplex DNA, passing another strand through the nick and changing the linking number by steps of one. Type II topoisomerases act by transiently nicking both strands of the DNA, passing another double stranded DNA segment through the gap and changing the linking number in steps of two with the help of ATP molecules.[3] A topoisomerase reaction has three general mechanistic steps i.e

  i. Binding of an enzyme to the substrate DNA
  ii. Cleavage by trans-esterification reaction accompanied by the formation of a transient phosphodiester bond between a tyrosine residue in the protein and one of the ends of the broken strand and subsequent strand passage through the break, leading to change in the linking number
  iii. Strand religation and release of the enzyme as the DNA is religated.

Under normal condition, the covalent enzyme-DNA cleavable complexes are fleeting catalytic intermediates and are present in low steady state concentrations, which cells can tolerate. However, conditions that significantly decrease or increase the physiological concentrations of these breaks unleash a myriad of deleterious side effects, including mutations, insertions, deletions and chromosomal aberrations.[4] Thus all topoisomerases are fundamentally dualistic in nature, catalyzing essential cellular reactions and possessing an inherent dark side capable of inflicting great harm to the genome of an organism. For these reasons DNA topoisomerases have been recognized as potential chemotherapeutic targets for antitumour and antiparasitic agents.[5,6]

DNA topoisomerases can be classified into three evolutionary independent families: type IA, type IB and type II. The *Escherichia coli* topoisomerase I and topoisomerase III, *Saccharomyces cerevisiae* topoisomerase III and reverse gyrase belong to the type IA or type I-5' sub-family as the protein link is to a 5' phosphate in the DNA. The prototype of type IB or I-3' enzymes are found in all eukaryotes and also in vaccinia virus topoisomerase I where the protein is attached to a 3' phosphate.[3] Though essentially similar in their action, these enzymes have a broader specificity than that of *E. coli* enzyme. Despite the differences in the mechanism and specificity between the bacterial and eukaryotic enzymes, the yeast DNA topoisomerase I has been shown to functionally complement a bacteria mutant in DNA topoisomerase I.[7] A certain degree of divergence also exists in the substrate preference, cofactor requirement and subunit composition of different topoisomerase families. Type IA topoisomerases are able to relax only negatively supercoiled DNA and require magnesium and single-stranded stretch of DNA for their function. Topoisomerases IB, however, are able to relax both positively and negatively supercoiled DNA with equal efficiency and do not require a single-stranded region of DNA or metal ions for function.[8]

The type II family includes *E.coli* DNA gyrase, *E.coli* topoisomerase IV (par E), all known eukaryotic type II topoisomerases and archaic topoisomerase VI. Type II enzymes are homo dimeric (eukaryotic topoisomerase II) or tetrameric (gyrase), cleaving both strands of a duplex that changes in linking number in steps of two. The current mechanistic model for topoisomerase II catalysed reactions involves the binding of two segments of DNA: a G (gate) segment, which is cleaved in both strands by the enzyme with the formation of an ester bond between active tyrosines and 5'-phosphates in the DNA and a T (transport) segment, which is captured by an ATP operated clamp that passes through the enzyme-stabilized break in the G segment.[9,10]

The discovery of several new DNA topoisomerases has brought a deeper understanding of their important roles in living cells. The biological functions of DNA topoisomerases are deeply rooted in the double helical structure of DNA and the selection of double stranded DNA as substrate has set the stage for their entrance.[11] Broad classifications of the different types of topoisomerases in different organisms are represented in Table 1.

Because DNA topoisomerases play key roles in cellular processes affecting the topology and organization of intracellular DNA, it is important to define the physiological functions and understand the molecular basis of their action. Moreover, beyond their normal cellular activities, these enzymes are proven molecular targets for clinically useful anti-tumor[12-14] and anti-microbial drugs.[15-17] In this context work on topoisomerases from the parasites has been a growing focus of interest.

Toxic chemotherapy and increasing drug resistance of some parasite strains to classical drugs along with coinfection of *Leishmania* with HIV, have made them a severe threat to public health in developing countries. Development of vaccines is still under trial and improved therapy is desirable.

Table 1. Classification of type I and type II DNA topoisomerases from different species

Subfamily	Representative Members
IA	Bacterial DNA topoisomerase I & II
	Yeast DNA topoisomerase III
	DNA topoisomerase IIIα and III
	Mammalian DNA topoisomerase IIIα and III
IB	Vaccinia and Pox virus monomeric topo I
	Kinetoplastida bi-subunit topoisomerase I
	Mammalian mitochondrial topoisomerase I
	Eukaryotic monomeric topoisomerase I
IIA	Bacterial gyrase, DNA topoisomerase IV
	Phage T4 DNA topoisomerase
	Yeast DNA topoisomerase II
	Drosophila DNA topoisomerase II
	Mammalian DNA topoisomerase IIα and IIβ
IIB	*Sulfolobus shibate* DNA topoisomerase VI
	(subunit A homologous to yeast SPO11)

Topoisomerases of Kinetoplastid Parasites

Type I DNA Topoisomerase

Type I DNA topoisomerases were isolated from *L. donovani*,[18-19] *Trypanosoma cruzi*[20] and *Crithidia fasciculate*.[21] The purified active enzymes (65 - 79 kDa) were ATP- independent and found to be sensitive to topoisomerase I specific inhibitor, camptothecin.[21] Although immunolocalization studies for *C. fasciculata* topoisomerase I showed that it is situated in the nucleus rather than the kinetoplast,[21] it has been demonstrated in trypanosomes that camptothecin treatment induces kDNA minicircle cleavage.[22] This observation suggests a possible existence of topoisomerase I in the kinetoplast of trypanosomes.

The first DNA sequence of a topoisomerase I-like gene from the kinetoplastid, *L. donovani* was reported by Broccoli et al, 1999. The deduced amino acid sequence of this gene showed an extensive degree of homology with the central core DNA binding domain of other eukaryotic type IB topoisomerases, including several conserved motifs but having a variable C-terminus. The conserved active site motif SKXXY was absent in the deduced amino acids sequence. The over-expressed protein in *E. coli* failed to show any relaxation activity in vitro or complement a bacterial mutant deficient in topoisomerase I activity.[23]

Type IB enzymes are the sole targets for a class of anti-tumor agents, camptothecins.[24] Type IB activity has been purified from a number of kinetoplastids.[25] The difference in the sensitivity of kinetoplastid topoisomerase I for camptothecin,[26] prompted the search for topoisomerase I sequence from kinetoplastid parasites which uncovered the existence of unique topoisomerase I from these parasites.[27]

All eukaryotic type IB topoisomerases are monomeric and consist of four domains.[2] The unconserved amino terminal domain contains putative signals for nuclear localization of the enzyme and is highly sensitive to proteolysis and dispensable for in vitro activity.[28] The largest core domain is essential for enzyme activity and shows high phylogenic conservation, particularly in the amino acid residues interacting closely with DNA. The third domain is known as the linker, which is poorly conserved and is variable in length. Finally, the carboxy terminal domain is highly conserved and contains the SKINYL motif. Cleavage occurs by trans-esterification reaction involving nucleophilic attack by an active site tyrosine (Tyr 723 in human Topo I) on a DNA phosphodiester bond resulting in the formation of a covalent DNA 3' phosphotyrosyl linkage. In religation phase a similar trans-esterification reaction involves attack by the free DNA 5' hydroxyl that releases the enzyme from DNA.[29,30]

Starting from bacteria to human to viruses, topoisomerases I are encoded by a single gene that contains the highly conserved DNA-binding and catalytic domains on a single peptide. But in kinetoplastid parasites, topoisomerase I is encoded by two genes, which associate with each other to form a hetero-dimeric topoisomerase I enzyme within the parasite. Emergence of the bi-subunit topoisomerase I in the kinetoplastid family have brought a new twist in topoisomerase research related to evolution and functional conservation of type IB family. Genetic analyses identify a gene for a large subunit, namely LdTOPIL, on *L. donovani* chromosome 34, encoding for a 636-amino acid polypeptide with an estimated molecular mass of 73 kDa. This subunit is closely homologous to the core domain of human topoisomerase I. The gene for the small subunit LdTOP1S encoding a 262 amino acid polypeptide with a predicted molecular mass of 28kDa, in turn is found on the *L.donovani* chromosome 4. The small subunit contains the phylogenetically conserved "SKXXY" motif placed at the C-terminal domain of all type I DNA topoisomerases, which conserves a tyrosine residue playing role in DNA cleavage (Fig. 1A). LdTOPIL shows about 54% identity with core subdomain of human topoisomerase I but less than 22% identity with the linker and the C-terminal domain. On the other hand, LdTOPIS shows 43.5% sequence identity with the C-terminal domain of human topoisomerase I, including alignment of conserved sequences surrounding the catalytic tyrosine residue. LdTOPIL also deviates significantly from human topoisomerase I at loop regions bounded by LdTOPIL residues Pro62-His63, Asp114-His 118 and Pro 341-Asp 342

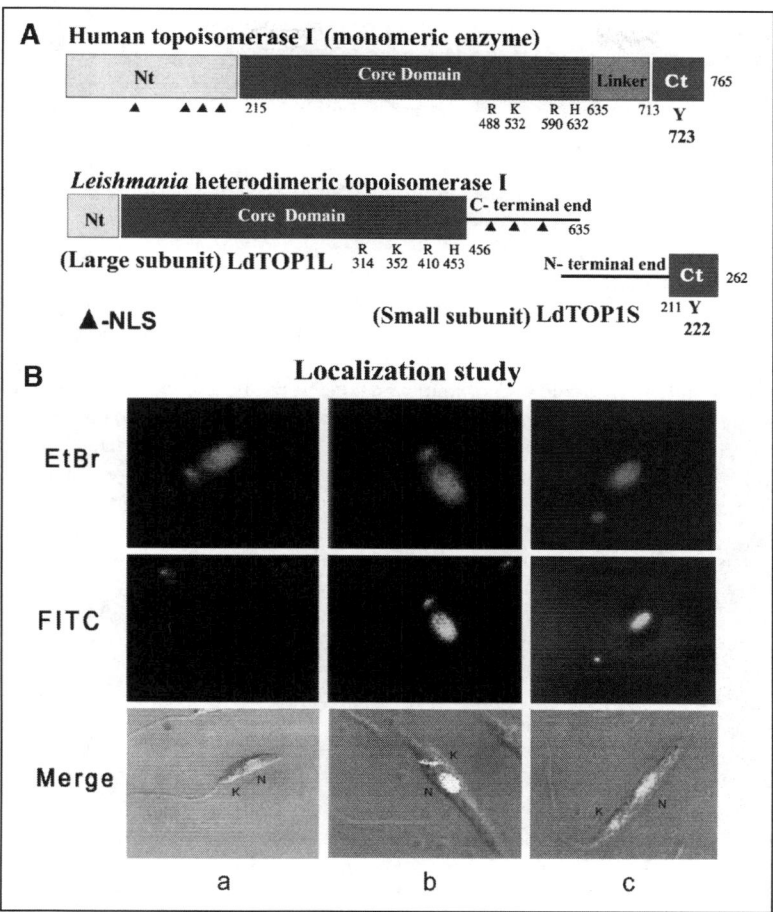

Figure 1. A) Schematic representation of the domain organization of monomeric human topoisomerase I and *Leishmania donovani* heterodimeric topoisomerase I. The domains are represented in different colored shades. Nt, N-terminal domain; Ct, C-terminal domain, NLS, Nuclear localization signal. Conserved residues are indicated in the figure. B) Immunocytochemical localisation with anti-LdTOP1L and LdTOP1S antiserum using fluorescent detection methods. Late log phase *L. donovani* promastigotes were fixed. No fluorescence was observed when preimmune serum was used as primary antibody and FITC-tagged secondary antibody (Panel a) as described by Das et al.[31] Panel b, same as Panel a, but probed with anti-LdTOP1L. Panel c, same as panel a, but probed with anti-LdTOP1S primary antibody. Parasite cells were also stained with ethidium bromide to locate the nucleus and kinetoplast and the area of the overlapping FITC and ethidium bromide (EtBr) stain are shown in merged pictures. Cells were viewed at an original magnification of 100 X under a Leica DM IRB inverted microscope. The nucleus (N) and kinetoplast (K) are indicated. Reproduced from reference 31.

which do not share the conserved sequences. Overall, this similarity indicates that the structure and catalytic machinery of the two enzymes are highly conserved, despite the fact that one is monomer and other is heterodimer (Fig. 1A).

Das et al,[31] described for the first time the in vitro reconstitution of the two recombinant proteins LdTOP1L and LdTOP1S corresponding to the large and small subunits. The proteins were purified from bacterial extract and the activity was measured by plasmid DNA relaxation assay. LdTOP1L and LdTOP1S forms a direct 1:1 heterodimer complex through protein-protein interaction. Under standard relaxation assay condition (50 mM KCl and 10 mM Mg^{2+}) reconstituted enzyme (LdTOP1LS) showed reduced processivity as well as 2 fold reduced affinity for DNA compared to eukaryotic monomeric rat liver topoisomerase I. Cleavage assay at various salt concentrations reveal that camptothecin (CPT) enhanced the formation of "cleavable complex" at low salt. Interaction between the two subunits leading to the formation of an active complex could be explored as an insight for development of new therapeutic agents with specific selectivity.

This observation leads to the concept that, non covalent interaction of both subunits is necessary for the activity. This was further evidenced from the charge difference of the two subunits. LdTOP1L has pI of 9.47 while that of LdTOP1S is 5.27. This charge difference clearly shows that these individual subunits are unstable until they interact with one another in the presence of salt. Recent findings by Das et al,[32] reveal that deletion of 99 amino acids from the N-terminus of LdTOP1L results in a protein which failed to interact with the smaller subunit. This could be attributed to the presence of many polar residues in this region. Polar interactions are common between the subunits of heterocomplex proteins. The overall charge difference for the large and small subunits in conjunction with the unusual salt sensitivity of the parasite protein suggests that ionic interactions are important for holding the subunits together.

Moreover it was established that silencing of one subunit in *T. bruci* causes the coordinate loss of both subunits of DNA topoisomerase I as well as results in a rapid reduction in the synthesis of both DNA and RNA of kinetoplastid parasites.[33]

Das et al,[34] also reveals that deletion of 39 aminoacids from the N-terminus of LdTOP1L results in a protein with decreased cleavage activity and sensitivity to CPT. These data argued in the favor of the interpretation that N-terminal amino acids of the large subunit regulates DNA dynamics during relaxation by controlling noncovalent DNA-binding or by coordinating DNA contacts by the other parts of the enzyme.

Davies and his co workers[34] have made a 2.27Å crystal structure of an active truncated *L. donovani* TOP1L/TOP1S heterodimer bound to nicked double stranded DNA in the presence of vanadate. The vanadate forms covalent linkages between the catalytic tyrosine residue of the small subunit and the nicked ends of the scissile DNA strand. This study reveals that arginine 410 residue of LdTOP1L (Arg 590 in human topoisomerase I) activates tyrosine 222 residue of LdTOP1S (Tyr 723 in human topoisomerase I) for attack on the scissile phosphate group, with water acting as a specific base. Moreover, it was also observed that Lys 352 of LdTOP1L (Lys 532 in human topoisomerase I) acts as the general acid in the cleavage reaction. Comparison of LdTOP1LS to the structure of human topoisomerase I bound to DNA containing topotecan reveals that all of the amino acids that form the drug binding pocket are completely conserved between the two species (Fig. 1A).

Das et al, showed that LdTOP1LS localize in both nucleus and kinetoplast of *L. donovani* (Fig. 1B). The existence of multiple localization signals have been mapped in the larger subunits of *Trypanosoma* and *Leishmania* topoisomerase I[25,27] but no NLS has been found in smaller subunits of the enzyme. So it is likely that the subunits interact in the cytosol before nuclear and kinetoplast importation. But, whether the proteins perform separate functions in the cytoplasm is still unknown.

Type II DNA Topoisomerase

Topoisomerase II activities have been purified from various kinetoplastid parasites.[25] Topoisomerase II genes have also been cloned from *C. fasciculata, T. brucei, T. cruzi, L. donovani, L. infantum, L. chagasi,* and *Bodo saltans*.[25] The genes and proteins of the parasites were found to be smaller compared to higher eukaryotes. No gyrase like activity (capable of introducing supercoils into DNA) has been found and the enzymatic activities and the genes are more like other eukaryotic counterparts. The topoisomerase activity isolated from *C. fasciculata* was shown to be immunolocalized in kinetoplast[35] but the overexpressed proteins from *L. donovani* and *B. saltans* were found to be localized both in the nucleus and kinetoplast.[36,37] Although it can be argued that the differences in cellular localization might be explained in terms of which epitopes were available for the recognition by different anti sera none the less the existence of another topoisomerase II sequence (hypothesized to be a mitochondrial topoisomerase II) cannot be dismissed. It is likely that replication of catenated kinetoplast DNA requires another topoisomerase activity.

Though all type II A topoisomerases are identical in one way that they change the linking number of DNA in an ATP-dependent manner, the eukaryotic type II enzymes are homodimers, while their bacterial counterparts like gyrase and topo IV are A_2B_2 tetramers; the B and A subunits being the N and C-terminal halves of their eukaryotic counterparts.[38] Certain phage type II topoisomerases are $A_2B_2C_2$ hexamers but they share the functional domains with other type II enzymes.[39,40] Interest in parasite type II topoisomerases and their genes and proteins gains impetus from the fact that they are the key enzymes involved in replication of the massive kinetoplast DNA network and RNAi of topoisomerase II leads to the progressive degradation of mitochondria in *Trypanosoma* . Kinetoplastid parasites diverged early in the eukaryotic evolution at the base of the evolutionary tree well before the emergence of kingdom metazoa. Inspite of having a similarity and identity of 31% and 23% with yeast topoisomerase II, LdTOP2 was found to complement a temperature sensitive mutant yeast strain.[41]

Just like other eukaryotic topoisomerase II, the parasite enzyme can also be divided into an N-terminal ATPase, a central DNA-binding and an unconserved C-terminal domain.[41-43] In spite of being unconserved, the nuclear localization signal and the dimerization domain of this homodimeric enzyme have been mapped in the C-terminus.[41] The C-terminus also contains a stretch of 60 amino acids not present in the human host. Therefore this region can be exploited to develop anti-leishmanial targets. The parasite enzyme has a greater affinity for DNA and was also stable at a very high salt concentration as compared to its human host.[42] These findings were quite consistent with the greater susceptibility of the parasite protein to the anti-topoisomerase II agents. This is because of the fact that an enzyme with more affinity towards DNA would perform more DNA cleavage and thus a greater chance of being trapped in that state by an anti-topoisomerase II drug.

The N-terminal 385 amino acids residues of LdTOP2 were found to possess the ATPase activity. Although the ATPase activity resides in the first 385 amino acid residues, only a larger protein was found to mimic the full-length enzyme kinetics in in vitro assay.[43] The study identifies specific amino acids like Asn65, Asn69, Asn96 and Asp130 of the parasite protein that are involved in the interaction with ATP and etoposide. In contrast, the ATPase domain of human topoisomerase IIα (1-453 amino acids) displays similar catalytic properties, in terms of ATP turnover, to that of the full-length enzyme, except for the fact that the smaller fragment (1-420 amino acids) fails to be hyperstimulated by DNA.[44] Most interestingly the ATPase activity of the N-terminal 385 amino acids of the parasite protein was also found to be inhibited by etoposide. Thus etoposide, in addition to being a poison for the parasite enzyme[42,43] is also a catalytic inhibitor of the enzyme.[43] The active site tyrosine implicated in DNA breakage and rejoining for *L. donovani* topoisomerase II has been mapped to be Tyr[775].[42] This tyrosine is the only residue in the parasite protein, which is involved in the trans-esterification reaction and is also homologous to the Tyr[804] of human.[45] Surprisingly, the C-terminal truncation mutants of the parasite protein fail to be inhibited by etoposide[42] compared to the full length enzyme. Like the human enzyme, the core domain of LdTOP2 contains all the elements essential for sequence preference in protein-DNA interaction, but unlike the human enzyme, the C-terminus of the parasite protein plays an important role in the in vitro topoisomerase II cleavage reaction.

It was observed earlier that over expression of human N-terminal domain in yeast confers resistance to high concentrations of etoposide.[46] The observed phenotype was proposed to be due to the competition of the excess of the N-terminal domain with the full length enzyme for a limiting pool of inhibitor. So future challenge in the parasite topoisomerase II would be to develop drug resistant parasite strains and to see what causes this resistance and also to check what effect the individual domains of the enzyme have on the drug protein interaction in the context of the full-length enzyme.

Topoisomerases as Therapeutic Targets

Despite differences in catalytic mechanism and cellular functions, the critical feature of all topoisomerases is the DNA strand passage event. However, the ability to pass single or double-stranded segment of DNA freely through another comes with a heavy price; it requires enzymes that generate breaks in the genetic material. In an effort to maintain genomic integrity during this cleavage reaction, topoisomerases covalently attach to the newly generated DNA 3' (eukaryotic topoisomerase I) or 5' termini (all other topoisomerases via phosphotyrosyl bonds. Under normal circumstances, these covalent enzyme-DNA cleavage complexes are transient catalytic intermediates and are present in low concentrations and consequently, they are tolerated by the cell. However, conditions that significantly increase the physiological concentrations cause deleterious side effects, including mutations, insertions, deletions and chromosomal aberrations. Thus, all topoisomerases are fundamentally dualistic in nature. Although they catalyze essential reactions in the cell, they possess an inherent dark side capable of inflicting great harm to the genome of an organism.

Classification of Topoisomerase Inhibitors

Topoisomerase-targeting therapeutics currently in use act by trapping the covalent enzyme-DNA complexes of the first trans-esterification reaction. The known topoisomerase drugs can be divided into two classes, class I and class II.[5,47] The class I drugs have been referred to as 'topoisomerase poison' where as the class two drugs are referred to as 'topoisomerase inhibitors'. The class I drugs act by stabilizing the topoisomerase-DNA covalent complexes. These include bacterial gyrase inhibitors quinolones, eukaryotic topoisomerase I inhibitor camptothecin and topoisomerase II inhibitors amsacrine, doxorubicin, etoposide and teniposide. The class II drugs interfere with catalytic function of DNA topoisomerase without trapping the covalent complexes (Fig. 2). These classes of drugs include the coumermycin family of antibiotics that act on bacterial gyrases, the eukaryotic DNA topoisomerase II inhibitor suramin, fostriecin, merbarone and bis-dioxopiperizines. Several inhibitors of eukaryotic topoisomerase I have also been reported.

A major determinant of cytotoxicity for the class I drug is the conversion of a latent single or double stranded break in a drug-topoisomerase-DNA complex into an irreversible double stranded break. Replication is the key cellular process that drives this conversion in case of the topoisomerase I drug camptothecin. However, for class II topoisomerase II drugs, processes other than replication might also be involved. Cell killing by class II topoisomerase II drugs may involve cell cycle progression through mitosis. Traversing of eukaryotic cells through mitosis in the absence of functional DNA topoisomerase II can lead to aneuploidy and chromosomal breakage. For class I drug, cytotoxicity increases with increasing cellular level of target enzyme where as for class II drugs opposite is true. Thus increased levels of topoisomerases render cells hypersensitive to enzyme poisons but resistant to inhibitors. Conversely, decreased enzyme levels render cells resistant to poison but hypersensitive to inhibitors.

Topoisomerases as Targets for Antiparasitic Agents

Sodium stibogluconate and ureastibamine, the two most potent and therapeutically used antileishmanial drugs have been reported by this laboratory to be specific inhibitors of *L. donovani* DNA topoisomerase I.[48] Pentavalent antimonials, also used as antileishmanial drugs, have been found to stabilize cleavable complex with an ED_{50} of 16.7 μg / ml and 209.5 μg / ml for wild type and resistant strains respectively.[49]

CPT, a plant alkaloid, an important class of antitumor agent[5] represents the best characterized topoisomerase IB inhibitor. It is reported to inhibit DNA topoisomerase I of *Leishmania* and *Trypanosoma*.[25] CPT is an uncompetitive inhibitor that directly traps the topoisomerase I-DNA covalent complex and slows the religation step of the nicking closing cycle.[2] CPT also hinders or blocks DNA rotation, which is evidenced by the crystal structure of the ternary complex between human topoisomerase I (topo 70) covalently linked to the DNA and the

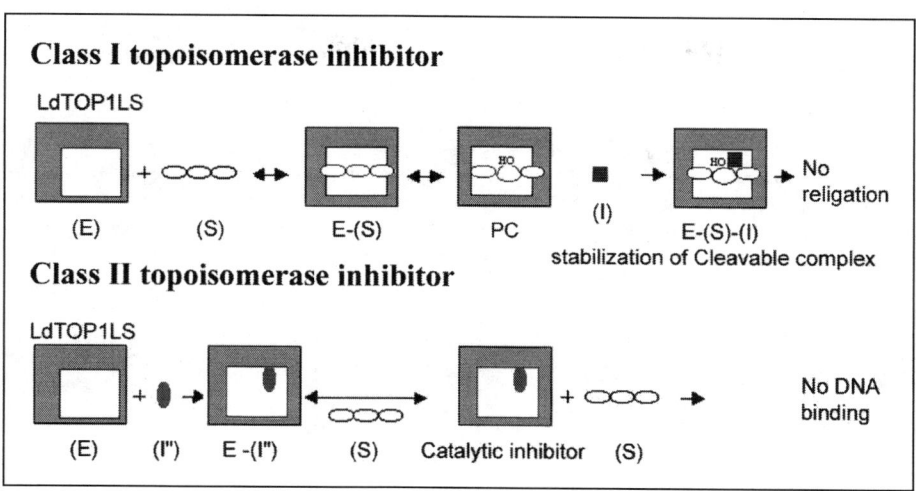

Class I topoisomerase inhibitor

LdTOP1LS

(E)   (S)   E-(S)   PC   (I)   E-(S)-(I) → No religation

stabilization of Cleavable complex

Class II topoisomerase inhibitor

LdTOP1LS

(E)   (I″)   E -(I″)   (S)   Catalytic inhibitor   (S) → No DNA binding

Figure 2. Schematic representation of mechanism of inhibition of bi-subunit topoisomerase I. Catalytic cycle of topoisomerase I is divided into DNA binding, cleavage and religation. A) CPT, a Class1 inhibitor, binds to the enzyme-DNA post-cleavage complex (PC), and subsequently inhibits religation step, and thus stabilizes the catalytic intermediate, (B) DHBA a pentacyclic triterpenoid, a Class II inhibitor, binds in the catalytic site of the enzyme and prevents binding of the enzyme with the DNA. E- is reconstituted bi-subunit *L. donovani* topoisomerase I (LdTOP1LS), where the green box represents the DNA binding large subunit (LdTOP1L) while the white small box is the catalytic subunit (LdTOP1S) harbouring SKXXY motif, S- substrate DNA, E-(S)- enzyme substrate complex, (I)- CPT, (I″)-DHBA. A color version of this figure is available online at www.eurekah.com.

CPT derivative topotecan.[2] Recent finding reveals that a highly CPT resistant *L. donovani* strain (LdRCPT.160) developed by stepwise exposure to CPT induces point mutations (Gly 185 Arg and Asp325 Glu) in the large subunit (LdTOP1L) of the bi-subunit topoisomerase I. The mutant enzyme shows reduced activity as well as reduced sensitivity towards CPT.[50] The cytotoxicity of 9-substituted-10, 11-methylenedioxy analogs of camptothecin correlate well with cleavable complex formation in the nucleus and kinetoplast, and structural motifs have been identified that disproportionately increase toxicity to parasites, compared with mammalian cells. Sen et al,[51,52] has demonstrated that CPT induces programmed cell death (PCD) both in the amastigotes and promastigotes form of *L. donovani* parasite.

Structure-activity relationship studies with mitonafide have revealed that the compound inhibits both nuclear and mitochondrial topoisomerase of *Leishmania* with preferential targeting of the mitochondrial enzyme over the nuclear enzyme.[53] Anilinoacridines have recently been found to possess antiparasitic activity towards *Leishmania, Trypanosoma* and *Plasmodium* species. These compounds have been shown to induce protein associated DNA lesions in *L. chagasi* promastigotes. Linearization of kinteoplast DNA minicircles have also been reported in parasites treated with anilinoacridines at similar concentrations.[54] Members of the 9-anilinoacridine topoisomerase II inhibitors have also been shown to inhibit growth of *L. major* promastigotes and amastigotes.[55] 9-aminoacridines, that are reported topoisomerase II inhibitors and structurally related to the antileishmanial compound quinacrine and chlorpromazine, have shown anti leishmanial activity at concentrations in the range of 10-20 μM.[56]

For the last decade, our laboratory has been involved in the search of DNA topoisomerase targeted novel anti-leishmanial agents from various indigenous plants. Towards this goal we have isolated some compounds with profound antileishmanial effects. Amarogentin, isolated from *Swertia chirata*, was found to inhibit the catalytic activity of *L. donovani* DNA topoisomerase

I by preventing enzyme-DNA binary complex formation.[57] Administration of *L. donovani* infected golden hamsters with vesicular forms of amarogentin, liposome and niosomes, was found to be more effective than free amarogentin.[58] Indolyl quinoline, a biologically active synthetic compound, also acts as a dual inhibitor of *L. donovani* topoisomerase I and II.[59] Diospyrin, a bisnapthquinone isolated from *Diospyros montana,* have been reported to be a potent inhibitor of *Leishmania* topoisomerase I with no effect on topoisomerase II. Diospyrin requires a much higher concentration to inhibit calf thymus DNA topoisomerase I and has been reported to exhibit significant inhibitory effect on the growth of *L. donovani* promastigotes.[60] Chowdhury et al, reported that dihydrobetulinic acid (DHBA), a derivative of betulinic acid that exhibits anti-HIV activity, is another excellent inhibitor of *Leishmania* DNA topoisomerase I and II[61] with the potential to become a lead therapeutic compound.[61] DHBA is a potent anti-leishmanial agent that induces apoptosis by primarily targeting parasitic topoisomerases. The structure of potential inhibitors of *Leishmania* topoisomerases are shown in Figure 3.

We have shown that the flavonoids quercetin and luteolin, isolated from *Vitex nigundo,* have potent antileishmanial effect. The flavonoids inhibited the growth of *L. donovani*

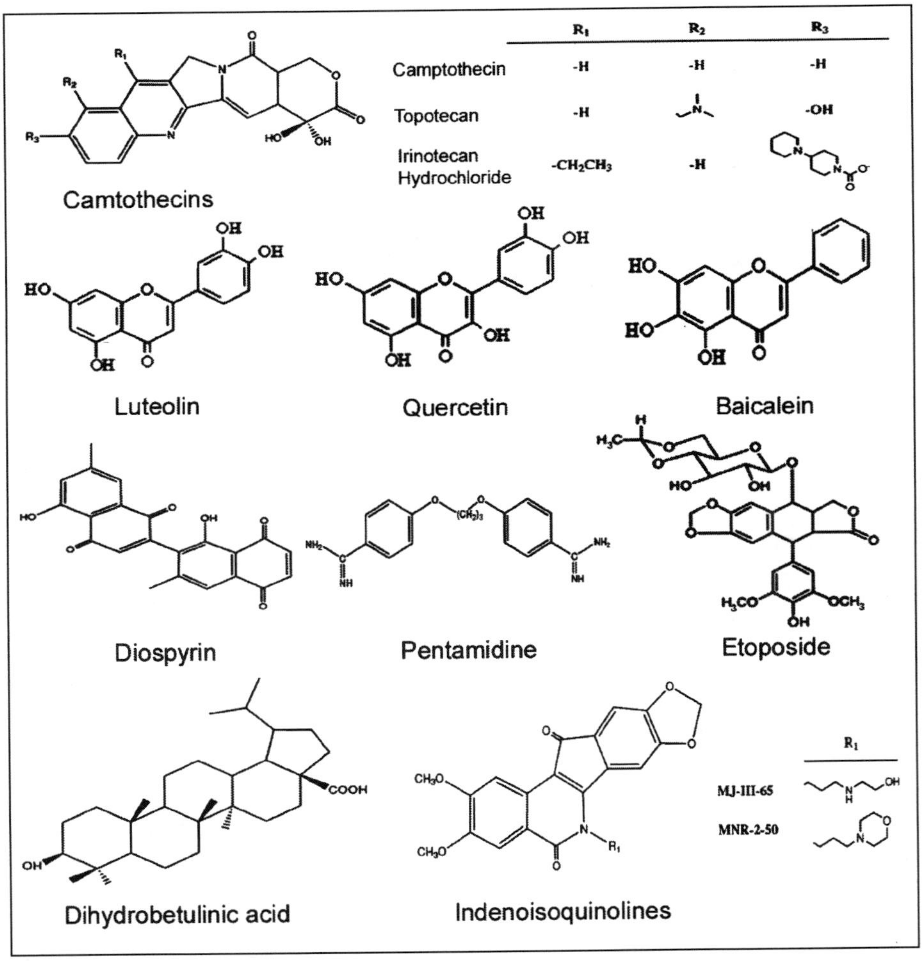

Figure 3. Structure of the potential *Leishmania* topoisomerases inhibitors.

promastigotes and amastigotes in vitro and also promoted topoisomerase II mediated linearization of kDNA minicircles. They arrested cell cycle progression in *L. donovani* promastigotes leading to apoptosis and reduced parasite burden in animal models.[62] Recently, Das et al,[63] described that naturally occurring flavones baicalein, luteolin and quercetin are potent inhibitors of the recombinant *Leishmania donovani* topoisomerase I. These compounds bind to the free enzyme and also intercalate into the DNA at a very high concentration (300 µM) without binding to the minor groove of DNA. The inhibition of topoisomerase I by these flavones is due to stabilization of topoisomerase I-DNA-cleavage complexes, which subsequently inhibit the religation step. Their ability to stabilize the covalent topoisomerase I-DNA complex in vitro and in living cells is similar to that of the known topoisomerase I inhibitor camptothecin (CPT). However, in contrast to CPT, baicalein and luteolin failed to inhibit the religation step when the drugs were added to preformed enzyme substrate binary complex. The most interesting part of the study reveals that baicalein and luteolin stabilize duplex oligonucleotide cleavage with CPT-resistant mutant enzyme LdTOP1Δ39LS lacking 1-39 amino acids of the large subunit.[32] This observation was further supported by the stabilization of in vivo cleavable complex by baicalein and luteolin with highly CPT-resistant *L. donovani* strain. Thus the interacting amino acid residues of *L. donovani* topoisomerase I may be partially overlapping or different for flavones and CPT.

Conclusion

Topoisomerase genes and proteins characterized from kinetoplastid parasite *Leishmania* appear to share many characteristics associated with their human homologues, but certain striking differences, including different enzyme activity requirements and different sensitivities to topoisomerase poisons provide insight for the development of topoisomerase-directed antiparasitic therapeutics. It has been established by several studies that the inhibitors of topoisomerases convert these essential enzymes into intracellular proliferating cell toxins and thereby provide a good tool for preferentially killing of the highly replicative parasite cells within the host. The interaction of the enzyme with specific inhibitors and poisons screened from natural or synthetic sources will help in the quest to selectively target the topoisomerase-based replication apparatus as a means to therapeutically control the parasitic menace in the foreseeable future.

Acknowledgements

This work was supported by the grants from Network Project SMM-003 of Council of Scientific and Industrial Research (CSIR), Government of India to H.K.M. Council of Scientific and Industrial Research (CSIR), Government of India supported B.B.D. and A.G with Senior Research Fellowships.

References

1. Croft SL, Sundar S, Fairlamb AH. Drug resistance in leishmaniasis. Clin Microbiol Rev 2006; 19:111-26.
2. Wang JC. Cellular roles of DNA topoisomerases: A molecular perspective. Nat Rev Mol Cell Biol 2002; 6:430-40.
3. Champoux JJ. DNA topoisomerases: Structure, function, and mechanism. Annu Rev Biochem 2001; 70:369-413.
4. Ferguson LR, Baguley BC. Topoisomerase II enzymes and mutagenicity. Environ Mol Mutagen 1994; 24:245-261.
5. Liu LF. DNA topoisomerase poisons as antitumor drugs. Annu Rev Biochem 1989; 58:351-75.
6. Burri C, Bodley AL, Shapiro TA. Topoisomerases in kinetoplastids. Parasitol Today 1996; 6:226-31.
7. Bjornsti MA, Wang JC. Expression of yeast DNA topoisomerase I can complement a conditional-lethal DNA topoisomerase I mutation in Escherichia coli. Proc Natl Acad Sci USA 1987; 84:8971-8975.
8. Stewart L, Ireton GC, Champoux JJ. Reconstitution of human topoisomerase I by fragment complementation. J Mol Biol 1997; 269:355-372.

9. Berger JM. Structure of DNA topoisomerases. Biochim Biophys Acta 1998; 1400:3-18.
10. Bakshi RP, Galande S, Muniyappa K. Functional and regulatory characteristics of eukaryotic type II DNA topoisomerases. Crit Rev Biochem Mol Biol 2001; 36:1-37.
11. Wang JC. DNA topoisomerases: Why so many? J Biol Chem 1991; 266:6659-6662.
12. Broxterman HJ, Georgopapadakou N. Cancer research 2001: Drug resistance, new targets and drug combinations. Drug Resist Updat 2001; 4:197-209.
13. Nitiss JL. DNA topoisomerases in cancer chemotherapy: Using enzymes to generate selective DNA damage. Curr Opin Investig Drugs 2002; 3:1512-1516.
14. Denny WA, Baguley BC. Dual Topoisomerase I/II Inhibitors in Cancer Therapy Curr. Top Med Chem 2003; 3:339-353.
15. Shapiro TA. Inhibition of topoisomerases in African trypanosomes. Acta Trop 1993; 54:251-260.
16. Bodley AL, Wani MC, Wall ME et al. Antitrypanosomal activity of camptothecin analogs. Structure-activity correlations. Biochem Pharmacol 1995; 50:937-942.
17. Bearden DT, Danziger LH. Mechanism of action of and resistance to quinolones. Pharmacotherapy 2001; 21:224S-232S.
18. Chakraborty AK, Majumder HK. Mode of action of pentavalent antimonials: Specific inhibition of type I DNA topoisomerase of Leishmania donovani. Biochem Biophys Res Commun 1988; 152:605-611.
19. Chakraborty AK et al. A type I DNA topoisomerase from the kinetoplast hemoflagellate Leishmania donovani. Ind J Biochem Biophys 1993; 30:257-263.
20. Riou GF et al. A type I DNA topoisomerase from Trypanosoma cruzi. Eur J Biochem 1983; 134:479-484.
21. Melendy T, Ray DS. Purification and nuclear localization of type I topoisomerase from Crithidia fasciculata. Mol Biochem Parasitol 1987; 24:215-225.
22. Bodley AL, Shapiro TA. Molecular and cytotoxic effects of camptothecin, a topoisomerase I inhibitor, on trypanosomes and Leishmania. Proc Natl Acad Sci USA 1995; 92:3726-3730.
23. Broccoli S, Marquis JF, Papadopoulou B et al. Characterization of a Leishmania donovani gene encoding a protein that resembles a type IB topoisomerase. Nucleic Acids Res 1999; 27:2745-52.
24. Champoux JJ. Domains of human topoisomerase I and associated functions. Prog Nucleic Acid Res Mol Biol 1998; 60:111-32.
25. Das A, Dasgupta A, Sengupta T et al. Topoisomerases of kinetoplastid parasites as potential chemotherapeutic targets. Trends Parasitol 2004; 8:381-387.
26. Bodley AL, Shapiro TA. Molecular and cytotoxic effects of camptothecin, a topoisomerase I inhibitor, on trypanosomes and Leishmania. Proc Natl Acad Sci USA 1995; 92:3726-3730.
27. Villa H, OteroMarcos AR, Reguera RM et al. A novel active DNA topoisomerase I in Leishmania donovani. J Biol Chem 2003; 278:3521-6.
28. Stewart L, Ireton GC, Champoux JJ. The domain organization of human Topoisomerase I. J Biol Chem 1996; 271:7602-8.
29. D'Arpa P, Machlin PS, Ratrie IIIrd H et al. cDNA cloning of human DNA Topoisomerase I: Catalytic activities of a 67.7kDa carboxy-terminal fragment. Proc Natl Acad Sci USA 1988; 85:2543-7.
30. Liu LF, Miller KG. Eukaryotic DNA topoisomerases: Two forms of type I DNA topoisomerases from HeLa cell nuclei. Proc Natl Acad Sci USA 1981; 78:3487-91.
31. Das BB, Sen N, Ganguly A et al. Reconstitution and functional characterization of the unusual bi-subunit type I DNA topoisomerase from Leishmania donovani. FEBS Lett 2004; 565:81-88.
32. Das BB, Sen N, Dasgupta SB et al. N-terminal region of the large subunit of Leishmania donovani bisubunit topoisomerase I is involved in DNA relaxation and interaction with the smaller subunit. J Biol Chem 2005; 280:16335-44.
33. Bakshi RP, Shapiro TA. RNA interference of Trypanosoma brucei topoisomerase IB: Both subunits are essential. Mol Biochem Parasitol 2004; 136:249-55.
34. Davies DR, Mushtaq A, Interthal H et al. The structure of the transition state of the heterodimeric topoisomerase I of Leishmania donovani as a vanadate complex with nicked DNA. J Mol Biol 2006; 357:1202-10.
35. Melendy T, Shelina C, Ray DS. Localization of a type II DNA topoisomerase to two sites at the periphery of the kinetoplast DNA of Crithidia fasciculata. Cell 1988; 55:1083-1088.
36. Das A, Dasgupta A, Sharma S et al. Characterisation of the gene encoding type II DNA topoisomerase from Leishmania donovani: A key molecular target in antileishmanial therapy. Nucleic Acids Res 2001; 29:1844-1851.
37. Gaziova I, Lukes J. Mitochondrial and nuclear localization of topoisomerase II in the flagellate Bodo saltans (Kinetoplastida), a species with noncatenated kinetoplast DNA. J Biol Chem 2003; 278:10900-10907.
38. Lynn R, Giaever G, Swanberg SL et al. Tandem regions of yeast DNA topoisomerase II share homology with different subunits of bacterial gyrase. Science 1986; 233:647-649.

39. Liu LF, Liu CC, Alberts BM. T4 DNA topoisomerase: A new ATP dependent enzyme for T4 bacteriophage DNA replication. Nature 1979; 281:456-461.
40. Seasholtz AF, Greenberg GR. Identification of bacteriophage T4 gene 60 product and a role for this protein in DNA topoisomerase. J Biol Chem 1983; 258:1221-1226.
41. Sengupta T, Mukherjee M, Mandal CN et al. Functional dissection of the C-terminal domain of type II DNA topoisomerase of kinetoplastid hemoflagellate Leishmania donovani. Nucleic Acids Res 2003; 31:5305-5316.
42. Sengupta T, Mukherjee M, Das R et al. characterization of the DNA-binding domain and identification of the active site residue in the gyr A half of Leishmania donovani topoisomerase II. Nucleic Acids Res 2005; 33:2364-73.
43. Sengupta T, Mukherjee M, Das A et al. Characterization of the ATPase activity of topoisomerase II from L. donovani and identification of the residues conferring resistance to etoposide. Biochem J 2005; 390:419-26.
44. Campbell S, Maxwell A. The ATP-operated clamp of human DNA topoisomerase IIalpha: Hyper-stimulation of ATPase by "piggy-back" binding. J Mol Biol 2002; 320:171-188.
45. Tsai-Pflugfelder M, Liu LF, Liu AA et al. Cloning and sequencing C- DNA encoding human topoisomerase II and localization of the gene to chromosome region 1988; 17q21-22.
46. Vilian N, Tsai-Pflugfelder M, Benoit A et al. Modulation of drug sensitivity in yeast cells by the ATP-binding domain of human DNA topoisomerase II alpha. Nucleic Acids Res 2003; 31:5714-5722.
47. Wang JC. DNA topoisomerase as targets of therapeutics: An overview. Adv Pharmacol 1994; 29A:1-9.
48. Chakraborty AK, Majumder HK. Mode of action of pentavalent antimonials: Specific inhibition of type I DNA topoisomerase of Leishmania donovani. Biochem Biophys Res Commun 1988; 152:605-11.
49. Lucumi A, Robledo S, Gama V et al. Sensitivity of Leishmania viannia panamensis to pentavalent antimony is correlated with the formation of cleavable DNA-protein complexes. Antimicrob Agents Chemother 1998; 42:1990-5.
50. Marquis JF, Hardy I, Olivier M. Topoisomerase I amino acid substitutions, Gly185Arg and Asp325Glu, confer camptothecin resistance in Leishmania donovani. Antimicrob Agents Chemother 2005; 49:1441-6.
51. Sen N, Das BB, Ganguly A et al. Camptothecin induced mitochondrial dysfunction leading to programmed cell death in unicellular hemoflagellate Leishmania donovani. Cell Death Differ 2004; 11:924-36.
52. Sen N, Das BB, Ganguly A et al. Camptothecin-induced imbalance in intracellular cation homeostasis regulates programmed cell death in unicellular hemoflagellate Leishmania donovani. J Biol Chem 2004; 279:52366-75.
53. Slunt KM, Grace JM, Macdonald TL et al. Effect of mitonafide analogs on topoisomerase II of Leishmania chagasi. Antimicrob Agents Chemother 1996; 40:706-709.
54. Werbovetz KA, Spoors PG, Pearson RD et al. Cleavable complex formation in Leishmania chagasi treated with anilinoacridines. Mol Biochem Parasitol 1994; 65:1-10.
55. Gamage SA, Figgitt DP, Wojcik SJ et al. Structure-activity relationships for the antileishmanial and antitrypanosomal activities of 1'-substituted 9-anilinoacridines. J Med Chem 1997; 40:2634-42.
56. Werbovetz KA, Lehnert EK, Macdonald TL et al. Cytotoxicity of acridine compounds for Leishmania promastigotes in vitro. Antimicrob Agents Chemother 1992; 36:495-497.
57. Ray S, Majumder HK, Chakravarty AK et al. Amarogentin, a naturally occurring secoiridoid glycoside and a newly recognized inhibitor of topoisomerase I from Leishmania donovani. J Natl Prod 1996; 59:27-29.
58. Medda S, Mukhopadhyay S, Basu MK. Evaluation of the in-vivo activity and toxicity of amarogentin, an antileishmanial agent, in both liposomal and niosomal forms. J Antimicrob Chemother 1999; 44:791-4.
59. Ray S, Sadhukhan PK, Mandal NB et al. Dual inhibition of DNA topoisomerases of Leishmania donovani by novel Indolyl quinolines. Biochem Biophys Res Commun 1997; 230:171-175.
60. Ray S, Hazra B, Mittra B et al. Diospyrin, a bisnaphthoquinone: A novel inhibitor of type I DNA topoisomerase of Leishmania donovani. Mol Pharmacol 1998; 54:994-999.
61. Chowdhury AR, Mandal S, Goswami A et al. Dihydrobetulinic acid induces apoptosis in Leishmania donovani by primarily targeting DNA topoisomerase I and II: Implications in antileishmanial therapy. Mol Med 2003; 9:26-36.
62. Mittra B, Saha H, Chowdhury AR et al. Luteolin, an abundant dietary component is a potent anti-leishmanial agent that acts by inducing topoisomerase II-mediated kinetoplast DNA cleavage leading to apoptosis. Mol Med 2000; 6:527-541.
63. Das BB, Sen N, Roy A et al. Differential induction of Leishmania donovani bi-subunit topoisomerase I-DNA cleavage complex by selected flavones and camptothecin: Activity of flavones against camptothecin-resistant topoisomerase I. Nucleic Acids Res 2006; 34:21-32.

CHAPTER 10

Antiparasitic Chemotherapy:
Tinkering with the Purine Salvage Pathway

Alok Kumar Datta,* Rupak Datta and Banibrata Sen

Summary

Distinguishable differences between infecting organisms and their respective hosts with respect to metabolism and macromolecular structure provide scopes for detailed characterization of target proteins and/or macromolecules as the focus for the development of selective inhibitors. In order to develop a rational approach to antiparasitic chemotherapy, finding differences in the biochemical pathways of the parasite with respect to the host it infects is therefore of primary importance. Like most parasitic protozoan, the genus *Leishmania* is an obligate auxotroph of purines and hence for requirement of purine bases depends on its own purine salvage pathways.

Among various purine acquisition routes used by the parasite, the pathway involved in assimilation of adenosine nucleotide is unique and differs significantly in the extracellular form of the parasite (promastigotes) from its corresponding intracellular form (amastigotes). Adenosine kinase (AdK) is the gateway enzyme of this pathway and displays stage-specific activity pattern. Therefore, understanding the catalytic mechanism of the enzyme, its structural complexities and mode of its regulation have emerged as one of the major areas of investigation. This review, in general, discusses possible strategies to validate several purine salvage enzymes as targets for chemotherapeutic manipulation with special reference to adenosine kinase of *Leishmania donovani*.

Systemic endotheliosis, commonly known as Kala-azar in India, is caused by the parasitic protozoon *Leishmania donovani*. The spread of leishmaniases follows the distribution of these vectors in the temperate, tropical and subtropical regions of the world leading to loss of thousands of human lives.[1] WHO has declared leishmaniasis among one of the six major diseases namely leishmaniasis, malaria, amoebiasis, filariasis, Chagas disease and schistosomiasis in its Special Programme for Research and Training in Tropical Diseases. Strategies for better prophylaxis and urgent therapies must be therefore devised to control this menace among poor and under privileged population. However, the possible availability of antiparasitic vaccines appears remote in near future. Therefore, chemotherapy remains the mainstay for the treatment of most parasitic diseases.

Selectivity of an antiparasitic compound must depend upon its mode of specific inhibition of parasite replication leaving host processes unaffected. In principle, these agents are expected to exert their selective actions against growth of the invading organisms by having one or both of the following properties:

  i. Selective activation of compounds in question by enzyme (s) from the invading organisms, which are not present in the uninfected cells.
  ii. Selective inhibition of vital enzyme(s), which are essential for replication of the parasites.

*Corresponding Author: Alok Kumar Datta—Emeritus Scientist, CSIR (Govt. of India), Indian Institute of Chemical Biology, 4, Raja S.C.Mullick Road, Jadavpur, Kolkata 700 032, India. Email: alokdatta@iicb.res.in

Drug Targets in Kinetoplastid Parasites, edited by Hemanta K. Majumder.
©2008 Landes Bioscience and Springer Science+Business Media.

In order to design specific compounds with the above characteristics, it is essential to have a thorough knowledge of the properties of the enzyme(s) and/or macromolecules which are unique to the parasite. Phylogenetic studies suggested that trypanosomatid parasites are relatively early-branching eukaryotic cells and indeed their cellular organization differs considerably from their mammalian hosts counterpart.[2] Various enzymes, metabolites or proteins identified in parasites and known to be absent from or strikingly different in the mammalian hosts were considered as ideal drug targets. Among the various metabolic pathways that are presently being studied for their prospects to be exploited as the target for chemotherapeutic manipulation, the most important are (i) purine salvage (ii) polyamine and thiol metabolism (iii) folate biosynthesis (iv) DNA replication (v) glycolytic and (vi) fatty acid biosynthetic pathways etc. A number of excellent reviews, describing the prospects and efficacies of these pathways, already exist in the literature.[3-5] Our laboratory is engaged in studying the pathways responsible for synthesis and assimilation of purine nucleotides in the parasitic protozoon *Leishmania donovani*. Therefore, we shall, for the constraint of space, try to restrict the discussion mostly with the purine salvage pathways of various *Leishmania* parasites with particular reference to the unique features of one of the enzymes of the purine salvage pathway viz AdK and its prospects as the chemotherapeutic target. However, contributions of other workers will also be discussed whenever essential and analogy will be drawn in order to make the reading coherent.

The *Leishmania* genus goes through a dimorphic life cycle.[3,4] It exists as a promastigote (extracellular form) in the sand fly vector but is converted to an amastigote (intracellular form) upon entry into mammalian macrophages. During this transformation process, the activities of a large number of proteins and/or enzymes have been reported to be stage-specifically altered and hence they could be prospective targets for development of chemotherapeutic regimen based on the exploitable differences of the parasitic proteins from their respective host counterpart.[6-12]

General Strategies for Development and Characterisation of Drug Targets in Trypanosomatids

The traditional approach to the development of new antiparasitic compounds consists of screening of a large number of compounds or extracts containing natural bio-active products against particular pathogens. This random approach is conducted without prior knowledge of either the molecular target(s) within the pathogen or the mode of action of the drugs. A large majority of antiparasitic compounds in use today were developed using these strategies. Subsequently, the molecular targets of few of these drugs viz. chloroguanide, pyrimethamine and trimethoprim, have been identified.[13]

However, the more contemporary and rational approach to drug discovery began with the identification of molecular targets within the parasite. Targets are selected based on their essential functions in the survival of the cell. In the recent years gene knockout approach is the routine procedure employed for studying a particular gene-function and has thus become a method of choice for target validation. Following selection and validation of the target protein, the objectives are to identify the details of their molecular structures and/or functions that can be exploited to design compounds inhibitory to the target molecule in question. The structure of the target can be studied by crystallisation of the protein or may be modelled using three-dimensional coordinates of amino acids from related or homologous protein whose crystal structure is known.[14] This structure-based approach to the design of compounds has benefits beyond the discovery of selective potent inhibitors as it provides an additional advantage for the prediction and resolution of drug-resistance problems. The process does not have to wait for the appearance of drug-resistant strains. Instead, possible drug-resistant strains can be generated in vitro using low concentrations of drugs. The development of resistance can then be studied by identifying possible mutations in the target proteins.[13,15] Subsequently, the mutations can be analysed using interactive computer graphic display systems, which visualises changes in the three-dimensional structure of the protein. These studies, apart from revealing the possible mechanism for altered drug binding, also may suggest ways for designing modified compounds for development of second generation of drugs for treatment of drug-resistant strains.

Because purine salvage pathway has been widely accepted as one of the plausible targets for the development chemotherapeutic regimen in most parasites including *L. donovani*, the subsequent sections will highlight the known complexities of this pathway and point out the differences observed during transformation of the parasite from its extracellular to its intracellular form.

Acquisition and Assimilation of Purines in *L. donovani* Promastigotes

It is now well established that all parasitic protozoa, including *L. donovani*, as opposed to most of the mammalian cells, lack the ability to synthesize purines de novo and thus use their own unique complement of salvage enzyme system to scavenge purines from the host.[16-19] Only nucleosides viz. adenosine, guanosine and inosine and their analogs viz. tubercidin, iodotubercidine etc. or nucleobases (adenine, xanthine and guanine) can be taken up by the cell surface transporters. The presence of two such specific cell surface nucleobase/nucleoside transporters (viz. LdNT1 and LdNT2) has been well documented for leishmania species.[20-22] They mediate the uptake of purine nucleosides as well as some purine analogs but with different specificities. However, host nucleotides must have to be converted into respective nucleosides prior to uptake. This task is accomplished by unique cell surface 3'-nucleotidase/nucleases.[23-26] The flow sheet diagram depicts the known pathways by which purine bases are scavenged and/or assimilated in *Leishmania* (Fig. 1). The key enzymes involved in this process

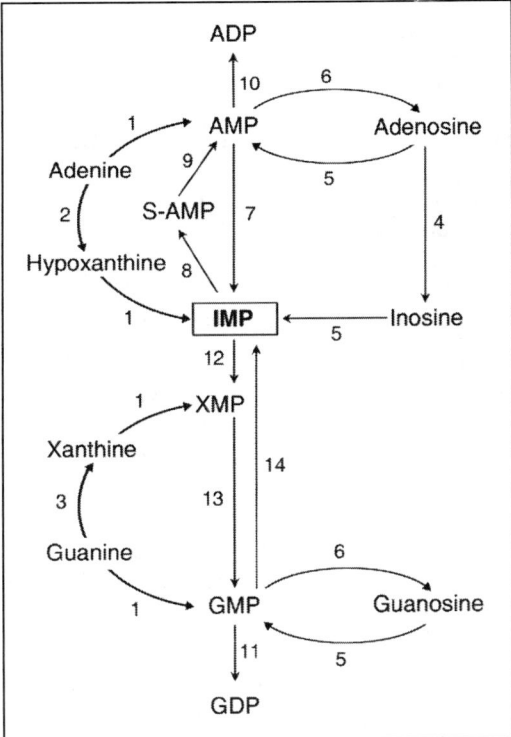

Figure 1. Purine salvage pathways of *Leishmania species*. Enzymes: 1) phosphoribosyltransferase; 2) adenine deaminase; 3) guanine deaminase; 4) adenosine deaminase; 5) nucleoside kinase; 6, nucleotidase; 7) AMP deaminase; 8) adenylosuccinate synthetase; 9) adenylosuccinate lyase; 10) AMP kinase; 11) GMP kinase; 12) IMP dehydrogenase; 13) GMP synthetase; 14) GMP reductase.

are adenine deaminase and guanine deaminase, which convert adenine and guanine to hypoxanthine and xanthine respectively. Interestingly, phosphorylase activity has not yet been detected in the *Leishmania*. Therefore, phosphoribosyl transferase (PRTase) activity appears to play a central role in the salvage of these purine bases.[27] Till date, three such PRTase activities viz. adenine phosphoribosyl transferase (APRTase), hypoxanthine-guanine phosphoribosyl transferase (HGPRTase) and xanthine phosphoribosyl transferase (XPRTase) have been identified and characterised in *Leishmania*.[27] These PRTases of the parasite then convert hypoxanthine and xanthine to inosine monophosphate (IMP) and xanthine monophosphate respectively. The enzyme XPRTase of *L. donovani* has been the focus of attention for a long time as it is absent in mammalian systems.[28] Moreover, since analogs of naturally occurring purine bases, known as subversive substrates, can function as prodrugs and enter nucleotide pool with lethal effect, this group of enzymes have stimulated considerable therapeutic interest with regard to a spectrum of parasitic diseases.

A stage-specific difference in the activities of enzymes is another characteristic of some *Leishmania* species, with promastigotes containing adenine deaminase and amastigotes containing adenosine deaminase. IMP formed in the cell can be converted to AMP by adenylosuccinate synthetase and adenylosuccinate lyase whereas XMP is converted to IMP by GMP synthetase and GMP reductase. Moreover, IMP dehydrogenase has also the ability to convert XMP to GMP.[27]

Nucleosides, following entry into the cell, are converted to mononucleotides by either phosphotransferases or by nucleoside kinases. Phosphotransferase activity has been detected in the extracts of *L. donovani* but these enzymes have been found to utilize only inosine analogs as the substrate.[29] Adenosine kinase (AdK), that directly phosphorylates adenosine (Ado) to AMP, is present both in promastigote and in amastigote of *L. donovani* and *L. mexicana mexicana* whereas guanosine kinase is present only in *L. m. mexicana*. Two other kinases viz. inosine and xanthosine kinases have also been detected in very low amounts. Adenosine deaminase, an important enzyme in mammalian cells, is however present in *Leishmania* amastigotes only.

Purine Metabolism in *L. donovani* Amastigotes

Purine metabolism in *L. donovani* and *L. m. mexicana* amastigotes has been extensively studied by Looker et al[8] and Hassan and Coombs.[30] The pathways of utilization of guanine, xanthine, hypoxanthine and their respective nucleosides are similar in both forms of the parasites. However, adenosine metabolism in *L. donovani* amastigotes differs markedly from that in promastigotes. In this connection, it may be mentioned that although a stage-specifically expressed adenosine transporter has been reported in amastigotes, confirmation of its existence is still awaiting.[31] In any case, following uptake in the amastigote, adenosine is deaminated to inosine by adenosine deaminase, not detectable in promastigotes. Subsequently, inosine is cleaved to hypoxanthine. However, adenine deaminase, that is known to deaminate adenine to hypoxanthine in the promastigote, is not present in the amastigote. In contrast, the activity of AdK in the amastigote shows 50-fold increase over the activity observed in the promastigote.[8]

It therefore appears that the *Leishmania* parasites possess multiple routes for salvaging purines and all the purine bases are interconvertible with an apparent branch point at IMP. Hence, the *Leishmania* species, unlike some other protozoa, when cultured in vitro, do not require any particular purine base for growth.

Purine Salvage Enzymes as Targets for Structure-Based Inhibitor Design

The potential of the purine phosphoribosyl transferases (APRTase, XPRTase, HGPRTase) and AdK as targets for antiparasitic chemotherapy stems from the major role of these enzymes in purine acquisition by the trypanosomatid parasites. However, because of the existence of various alternative purine salvage pathways (Fig. 1), it is conceivable that inhibition of a single enzyme would not kill the parasite. This has been confirmed by elaborate genetic investigation

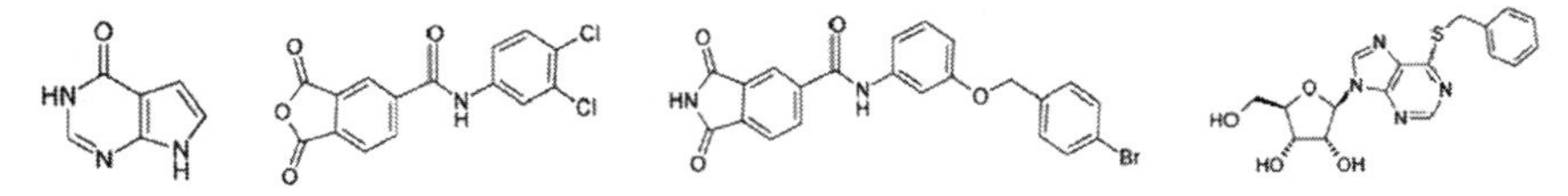

Figure 2. Chemical structures of various inhibitors of HGPRTase and AdK.

of *L. donovani* and *T. gondii*, suggesting that neither of these enzymes are essential for the parasite viability. In this connection it is to be noted that mutant *L. donovani* promastigotes lacking HGPRTase, APRTase and AdK, either singly or in any combination, can retain the capacity to proliferate in completely defined medium in which the sole exogenous purine source is any of the four naturally occurring nucleobases, hypoxanthine, xanthine, guanine or adenine or the nucleosides adenosine, inosine or guanosine, provided it has active XPRTase.[32] How knocking out of these genes will affect the survival of intracellular amastigotes is however remained to be seen. Therefore, it appears that a single chemotherapeutic agent or a combination chemotherapy targeting more than one enzyme would be ideal for successfully blocking the purine acquisition of the parasites.

Till date most of the structure-based inhibitor design strategies that target the purine salvage pathway of the parasite have been directed towards the HGPRTase. Owing to differences in the substrate specificity, the HGPRTase from *Leishmania* and trypanosomes,[33-35] in contrast to their host counterpart, phosphoribosylates antiparasitic pyrazolopyrimidines like allopurinol (Fig. 2), which subsequently is incorporated into the nucleic acids causing selective death of the parasite.[34,35] Allopurinol alone or in combination with other drugs, has been proved to be effective against cutaneous[36] and visceral leishmaniasis.[37] Solution of a number of apo, ion and product bound crystal structures of HGPRTase[38-40] reveals a closed active site that provides well-defined target for computational drug discovery. The structure-based docking method provided a remarkably efficient means for the identification of inhibitors targeting trypanosomal HGPRTase.[41] The inhibition constants of the lead inhibitors were very low with trypanostatic activity in cell culture.[41] Based on X-ray structure of *Tritrichomonas foetus* HGPRTase,[38] isatin and phthalic anhydride were also identified as two novel scaffolds capable of mimicking the substrate purine base and acting as competitive inhibitors of the target enzyme without any detectable effect on the human HGPRTase. TF1 (phthalic anhydride derivative) and TF2 (phthalimide derivative) (Fig. 2) were shown to be effective in killing the cultured parasites and the parasite growth inhibition could be reversed by addition of the natural substrate hypoxanthine to the culture medium.[42,43] These findings demonstrated the success of a structure-based computational approach whereby a molecule identified by computer-based means can be rationally modified to produce potent inhibitors of a chosen target enzyme and may provide useful starting point for drug design for the treatment of different parasitic diseases.

Prospects of Adenosine Kinase (AdK) as the Drug Target

Among the plethora of purine salvage enzymes in the trypanosomatid parasites, AdK is also being assessed as one of the potential chemotherapeutic targets for treating various parasitic diseases especially leishmaniasis[5] and toxoplasmosis.[5,44] Cohen et al[45] demonstrated that adenylate nucleotide pool is the major source of host purines in all mammalian cells. Since nucleotides do not enter cells readily, they are cleaved by different nucleotidases located on the surface of the parasites and adenine nucleoside is probably the first nucleoside to permeate the plasma membrane of the parasite,[23-25] which is then converted to its nucleotides and other nucleotides. Being the most active purine salvage enzyme, AdK reaction is the main route of adenosine metabolism in *T. gondii*.[46-48] This results in preferential incorporation of adenosine into adenine nucleotides by at least a 10-fold higher rate than that of any other purine precursor tested.[46,49] Mutant *L. donovani*, lacking AdK, incorporated 25% of the adenosine, indicating 75% of incorporated adenosine is directly phosphorylated by AdK in *L. donovani* promastigotes.[50] Another report indicated that, during transformation of promastigote to amastigote, AdK was stimulated almost 50-fold[8] and thus was suggested to play key role in the process.[51] A very recent study showed that AdK and HXGPRTase provide the only two physiologically relevant routes for purine acquisition in *T. gondii*. However, AdK knock-out parasites exhibited a greater fitness defect than HXGPRTase mutants, arguing that flux through AdK is greater than HXGPRTase.[52]

Structure-activity relationships as well as biochemical and metabolic studies established that the substrate specificity of *T. gondii* AdK differs significantly from those of the human enzyme. It was also demonstrated that AdK from *T. gondii*, as opposed to its host counterpart, preferentially metabolizes 6-benzylthioinosine (BTI) (Fig. 2) to the nucleotide level, which eventually acts as the toxic subversive substrate for the parasite.[5,44,53] Subsequently, several new classes of BTI analogues were synthesized by structure-based lead optimization, leading to further improvement of its antitoxoplasmic efficacy.[54,55] These findings are consistent with the notion that AdK indeed is a key purine salvage enzyme of the *Leishmania* and *Toxoplasma* species. Hence, understanding the reaction mechanism of AdK at the molecular level could be important both from a fundamental point of view as well as in the hope that detailed knowledge of the enzyme may provide relevant information necessary for designing specific inhibitors.

However, the parasitic AdK is one of the most under-exploited purine salvage enzymes as far as its structure-based inhibitor design is concerned. Lack of enough structural information about its active site and insufficient knowledge of the amino acid residues involved in the reaction mechanism have led to such an impasse. Recently however, the X-ray crystal structures of AdK from human and *T. gondii* have been solved.[56,57] Our laboratory has been working on the biochemistry of *L. donovani* AdK (LdAdK) for over two decades and has been a major contributor in unearthing various information regarding its reaction kinetics.[51,58,59] The enzyme from *L. donovani* has also been cloned and expressed, thereby providing workers in this field the necessary impetus to undertake structure-function analysis of the enzyme in a systematic manner.[60]

General Biochemical Properties of the *L. donovani* AdK

The enzyme from *L. donovani* is a 345-residue monomer of 38 KDa with pI of 8.8, sharply different from the pI (4.5-5.9) determined for AdK from other higher eukaryotic sources and is immunologically distinct from AdK of other sources.[51,58] The enzyme has a pH optimum of 7.5 and the activity is dependent on the optimum ATP-Mg2+ ratio. Studies showed that whereas the higher eukaryotic AdKs are prone to inhibition at high Ado and Mg^{2+} concentrations, LdAdK is refractory to such inhibition. In contrast, LdAdK is much more sensitive to inhibition by ATP.[51,59] However, despite these biochemical differences, the parasite enzyme, similar to other AdKs, is regulated by both its products, AMP and ADP.[59,61] LdAdK follows the typical sequential bi-substrate kinetics in which binding of Ado to the enzyme occurs prior to ATP binding with the release of AMP at the end.[59]

Structure of AdK from Different Sources

AdK sequences from mammalian sources show more than 90% amino acid identity among themselves.[62] Interestingly however, the translated amino acid sequences of enzymes from different sources bear no sequence similarity with other well-characterized nucleoside and nucleotide kinases, thus setting it apart from the family of other structurally and functionally related proteins. However, of particular interest is the two regions of AdK that has been found to be homologous with the members of the PfkB (phosphofructose kinase B) family of carbohydrate kinases viz ribokinase, inosine-guanosine kinase, fructokinase and 1-phosphofructokinase. Members of this family are characterized by the presence of two common sequence motifs that includes a highly conserved di-glycine motif located near the N-terminal end and a DTXGAGD motif, positioned near the C-terminus.[63,64]

Recently, the structure of AdK from human and *T. gondii* has been solved.[56,57] These high resolution structures were determined for the apo enzyme, AdK: Ado complex,[56,57] as well as the Ado: AMP-PCP (a nonhydrolysable ATP analogue) bound enzyme.[57] These findings revealed that the enzyme consists of two unequal-sized domains. The large domain is a three-layered sandwich of α helices and β sheets over which the small domain forms a lid.[65] The cleft formed between the two domains constitutes the catalytic site where several amino acids, probably responsible for Ado and ATP binding, are located. Although the ATP binding site of the enzyme from two sources is quite different, their Ado binding pockets are structurally similar. The location of the magnesium ion between the α and β phosphate of ATP is unusual and differs from several other kinases in which the cation is located between the β and γ phosphates. Nevertheless, the overall structure is similar to the reported structure of *E. coli* ribokinase, the first reported structure in the family of carbohydrate kinases.[66] Structures of human AdK and *E. coli* RK superimposed nicely with an RMS deviation of 2.4 Å, even though the sequence identity between them is only 22%.[56] In the overlapped structures, the ribose ligand of RK superimposed on the ribosyl group of Ado 1 and the adenosine portion of the ADP ligand in RK superimposes on Ado 2. This comparison provided strong evidence that Ado 1 exists in the binding site used for the nucleoside undergoing phosphorylation and that Ado 2 occupies the ATP/ADP-binding site. This was further unambiguously proved by analyzing the AMP-PCP bound structure of *T. gondii* AdK.[57]

Comparison of the structures of *T. gondii* AdK, in presence and in absence of the substrate further revealed a novel catalytic mechanism that involved both global and local changes in the protein structure upon binding of Ado. The most striking among them is the Ado-induced 30° hinge bending motion leading to domain closure. It was predicted that a GG conformational switch was responsible for this gross structural change that placed the enzyme in its precatalytic conformation.[57] Apart from these changes, other additional local structural changes were also shown to be induced by ATP binding. As a result of these transitions, an anion hole is created. In general, the most extensively characterized kinase anion hole is the mononucleotide binding motif or P-loop, which contains the consensus sequence GXX(G/X)XGK(S/T).[67] This motif has been observed in a number of enzymes that include adenylate kinase,[68,69] RecA,[70] elongation factor Tu[71] and p21ras[72] and also in the protein kinases.[73] But the AdK P-loop heptad, DTXGAGD, that encompasses the second ribokinase fingerprint region, is completely different and thus defined a new kinase anion hole motif.[57] Therefore, it is most likely that the synergistic substrate-induced structural changes lead to optimal juxtaposition of the substrates for the catalytic reaction between the adenosine 5'-hydroxyl oxygen atom and the ATP γ-phosphate. Both the structures indicated that the 5'-hydroxyl of adenosine is near to and in reasonable alignment with the γ-phosphate of ATP suggesting an in-line S_N2 displacement reaction.[56,57] Despite these structure-based predictions, the actual role of the active site residues involved either in the process of phosphate transfer or substrate binding remains to be biochemically validated.

Sequence Characteristics of LdAdK Gene and Homology Model-Based Structural Analysis of the Protein

Sequence comparison of the 5'-noncoding region of the AdK gene of *L. donovani* with its corresponding mRNA confirmed that, like other kinetoplastida genes, LdAdK transcript is processed post-transcriptionally following addition of the mini-exon at the 5' end of the mRNA. Furthermore, no consensus eukaryotic promoter sequences such as TATA or CCAAT could be identified upstream of the initiation codon,[74] a finding consistent with other kinetoplastida genes.

Homology alignment studies revealed that LdAdK has only about 40 and 31% identity with human and *T. gondii* enzymes respectively (Fig. 3A). However, despite this limited identity, LdAdK possesses all the characteristics typical of AdK from all known sources. First, similar to that of other AdKs, LdAdK lacks the consensus P-loop motif[60] and secondly, LdAdK harbors two amino acid sequence motifs that are distinctive of the PfkB family of carbohydrate

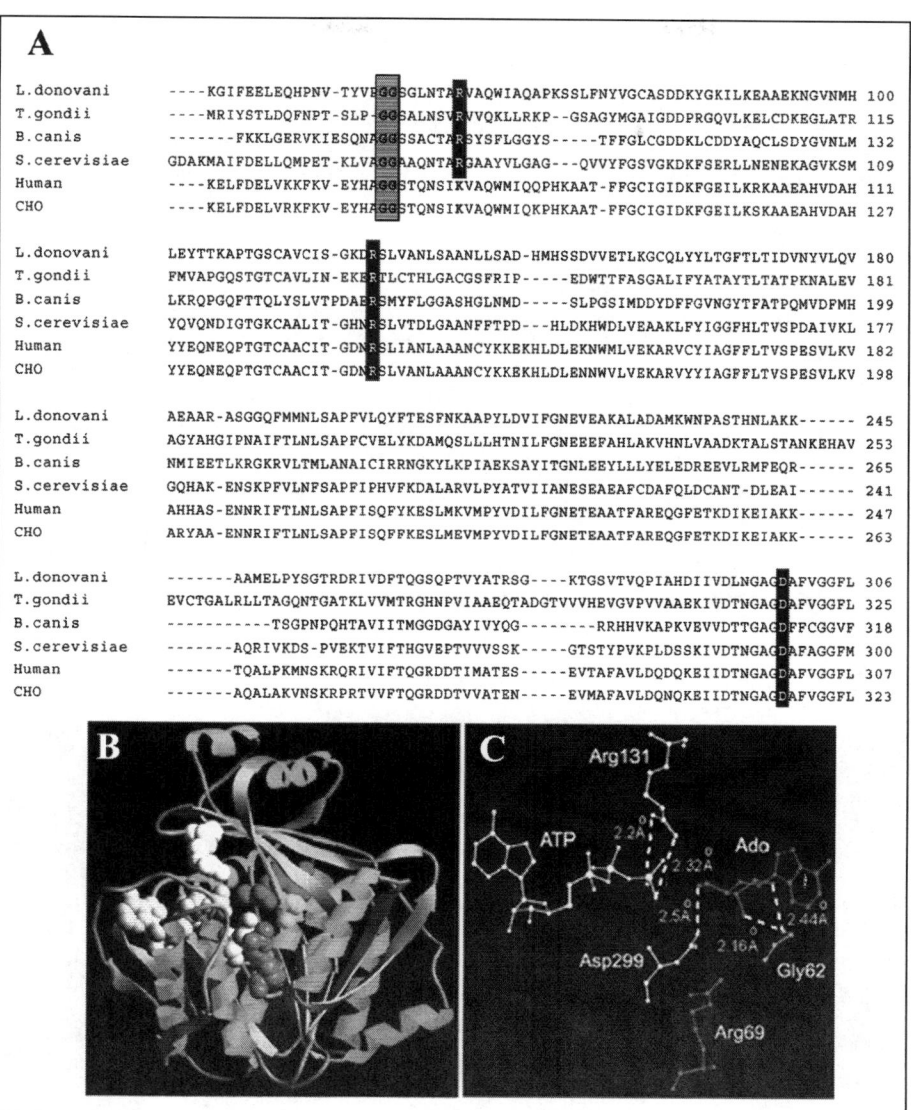

Figure 3. Amino acid sequence alignment and homology modeling of LdAdK. A) alignment of LdAdK amino acid sequence with that of *T. gondii* (AAF01261), *B. canis* (CAA11263), *S. cerevisiae* (P47143) human (AAA97893) and Chinese hamster ovary (AAA91649) AdK (numbers in the parenthesis indicate NCBI database accession numbers for respective proteins). Boxes indicate amino acid signature motifs distinctive of this family of proteins. The residues selected for site-directed mutagenesis i.e., Gly-62, Arg-69 and Arg-131 and Asp-299 of LdAdK are indicated in bold. B) homology model showing the overall structure of LdAdK with α-helices drawn as ribbons and β-strands as arrows. Spatial position of Adenosine (magenta), ATP (yellow), Gly-62 (cyan), Arg-69 (red), Arg-131 (white) and Asp-299 (pink) are shown in space-fill model. C) Zoomed picture of the active site residues with inter-atomic distances shown by dotted lines. Reproduced with permission from: Datta R et al, Biochem J 2005; 387:591-600; ©2005 The Biochemical Society.[78] A color version of this figure is available online at www.Eurekah.com.

kinases.[63,64] Studies further suggested that similar to other carbohydrate kinases, the Gly_{61}-Gly_{62} structural motif of LdAdK is probably essential for maintenance of its conformational flexibility. Of the seven arginine residues present in LdAdK, only Arg_{131}, corresponding to Arg_{136} and Arg_{132} of *T. gondi* and human AdK respectively, is absolutely conserved. Interestingly, another arginine residue, located at the 69th position of LdAdK seems to be conserved mostly among lower eukaryotes whereas in enzymes from higher eukaryotes this residue is replaced with lysine.[60,62,75-77] Among the acidic residues, Asp_{299} of LdAdK, corresponding to Asp_{318} and Asp_{300} of *T. gondi* and human AdK and located on the second fingerprint motif (DTXGAGD) of ribokinase family, was retained. Likewise, Asp_{16} of LdAdK was also strictly conserved with corresponding amino acid residues Asp_{24} and Asp_{18} of *T. gondii* and human AdK respectively.[65]

In the absence of experimentally determined structure of LdAdK, the three dimensional model, constructed on the basis of the sequence alignment and available coordinates from the human and *T.gondii* AdK crystal structures, showed a high level of overall structural and active-site geometrical symmetry among themselves (Fig. 3B). The ribbon diagram of the model shows that like AdK from other sources, LdAdK appears to be organized into two domains: one of the domains (large domain), consisting of a three-layered sandwich of ten α-helices and nine β-strands, is connected by four peptide segments to the smaller lid domain, consisting of five β-strands running perpendicular to two α-helices.

Identification of Potential Amino Acid Residues Involved in Catalysis

Taking advantage of the spatial coordinates of the model, the search for the amino acid residues within the interacting distance (≥ 3.5 Å) of the substrates, was made. In Table 1, residues of LdAdK that are within the interacting distance of either the base, Ado, or the ribose moiety of Ado and ATP or the phosphate group of ATP are listed.

From the analysis, the terminal phosphate group of ATP and the 5'-OH group of the ribose moiety of Ado appeared to be 1.68 Å, close enough for a direct in-line phosphate transfer. The structure further shows that Gly_{62} is located underneath the Ado-binding site of the peptide connecting β-4 sheet of the small domain with the α-3 helix of the large domain (Fig. 3C). Moreover, the peptide N of Gly_{62} is 2.44 Å and 2.16 Å away from the O2' and O3' group of the adenosyl ribose respectively, suggesting its possible role in Ado binding. Of the seven arginine residues present in the protein, only Arg_{131}, located on the β-8 sheet of the small domain, appeared to be spatially close to the active site. In fact, its NH1 and NH2 groups were found to be at potentially H-bonding distance of 2.2 Å and 2.3 Å respectively from the O2G and O3G groups of the terminal phosphate of ATP, an observation distinctly different from that of *T. gondii* AdK where, instead of two terminal amino

Table 1. Hydrogen bonds and close contacts (≤ 3.5 Å) between the purine or ribose subsite of adenosine and amino acid residues of LdAdK in the modeled structure

Purine		Å	Ribose		Å
N1	Asn12 ND2	2.81	O2'	Asp16 OD1	2.67
N1	Phe168 CB	3.11	O2'	Gly62 N	3.24
N1	Phe168 CG	3.28	O3'	Gly62 N	2.96
C2	Phe168 CD1	3.45	O3'	Gly62 CA	3.48
N3	Ser63 N	2.86	O3'	Gly62 C	3.25
C6	Phe168 CG	3.33	O3'	Asn66 ND2	2.94
N6	Phe168 CD2	3.44	C5'	Asn295 C	3.31
N7	Ala135 CB	3.40	C5'	Asn295 O	3.22
			O5'	Asp299 OD2	2.50

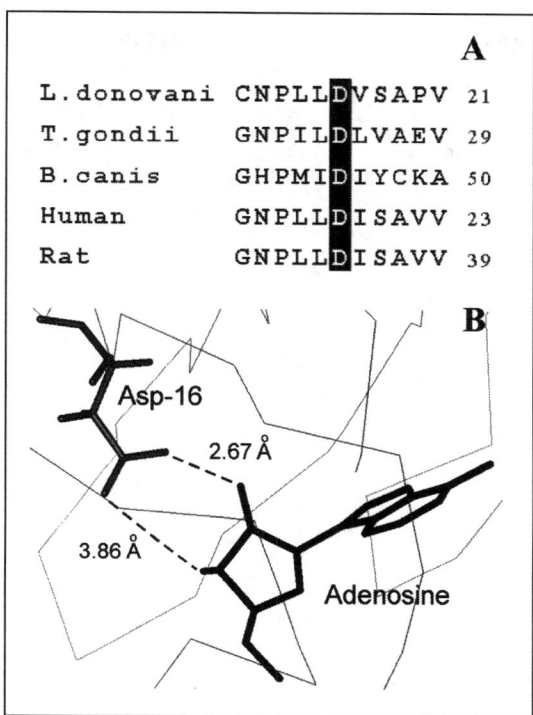

Figure 4. Sequence alignment and close-up view of the Ado-binding site of LdAdK. A) alignment of the N-terminal region of LdAdK amino acid sequence with that from *T.gondii* (AAF01261), *Babesia canis* (CAA11263), human (AAA97893), and rat (AAH81712) showed the invariant aspartate residue (numbers in parentheses indicate GenBank® accession numbers for respective proteins). B) structural model illustrating the position of Asp-16 relative to the bound Ado, dashed line depicts the possible interaction between the Asp-16 carboxyl group and ribose hydroxyls of Ado. Reproduced with permission from: Datta R et al, Biochem J 2006; 394:35-42; ©2006 The Biochemical Society.[63]

groups, NH1 and NH2 were postulated to be interacting with the terminal phosphate.[57] However, the spatial location of the Arg $_{69}$, located on the α-3 helix of the large domain, was found to be quite far away from either of the substrates (>10.0 Å). Apart from these residues, the carboxy side chain of Asp$_{16}$ and Asp$_{299}$ are also proximally located to the ribosyl O2′, O3′ and O5′ hydroxy group of Ado respectively (Figs. 3 and 4). Availability of these structural details allowed initiation of biochemical studies with regard to mechanisms of substrate binding, phosphate transfer and enzyme regulation.

Mechanism of Ado Binding

Crystal structures of human and *T. gondii* AdK demonstrated that the side chains of Asp$_{18}$ and Asp$_{24}$ respectively (the sequence homologous to Asp$_{16}$ of LdAdK) formed hydrogen bonds with both O2′ and O3′ ribose hydroxyls.[57] In the modelled LdAdK, its Asp$_{16}$ also points towards the adenosyl ribose (Fig. 4) and is proximal to its O2′ and O3′ groups. Moreover, comparison of the Ado-binding site of LdAdK and ribose binding site of ribokinase revealed that an Asp residue is conserved in both AdK and RK.[66] Structure-guided mutational analysis of the Asp$_{16}$ mutant demonstrated total obliteration of Ado binding to the enzyme, thereby indicating indispensability of the Asp$_{16}$ residue in Ado binding. Furthermore, possibility of the formation of a bidentate hydrogen bond between Asp$_{16}$ and the adenosyl ribose has also been proposed.[65]

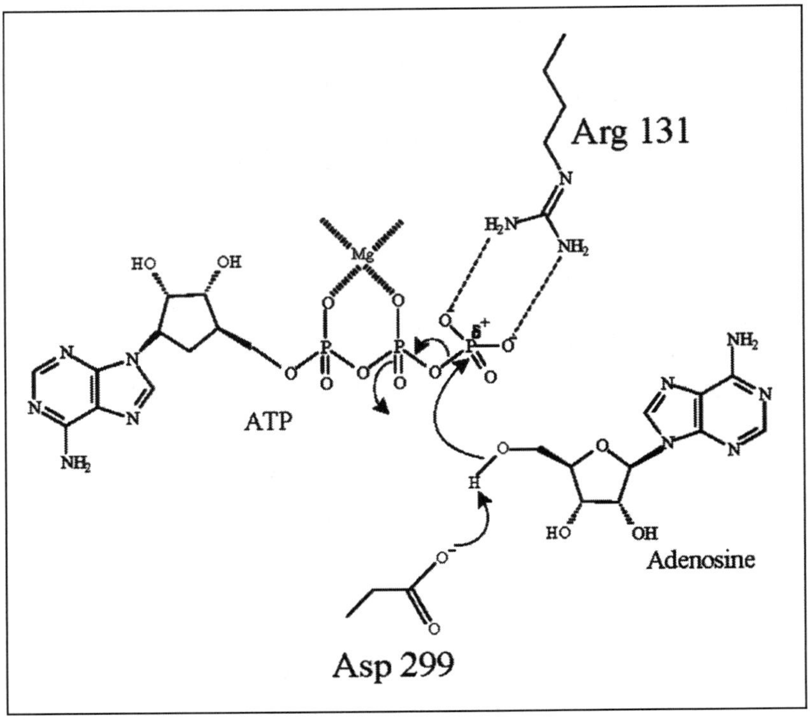

Figure 5. Schematic representation of the possible reaction mechanism of LdAdK; Ado-induced domain rotation around the flexible diglycine motif (Gly-61-Gly-62) places the enzyme in precatalytic conformation. ATP binding causes further conformational changes, resulting in the initiation of a series of events in which Asp-299 first withdraws a proton from the 5' hydroxy group of Ado (solid line) followed by a direct nucleophilic attack on the γ-phosphate of ATP (broken line). The resulting quinquivalent transition state is stabilized by Arg-131. Arg-131 also increases the electrophilicity (δ+) of the γ-phosphorus group. Reproduced with permission from: Datta R et al, Biochem J 2005; 387:591-600; ©2005 The Biochemical Society.[78]

Mechanism of Phosphate Transfer

Structural information and mutational analysis coupled with chemical modification of some of these residues led to development of a concerted mechanism for the phosphate transfer reaction[78] (Fig. 5). The mechanism suggests that initial binding of Ado to the open active site of the enzyme induces a domain rotation around the di-glycine hinge (Gly_{61}-Gly_{62}) motif. Arg_{69}, located on the α-3 helix possibly facilitates such domain movement. This leads to a relatively closed positioning of the lid and places the second substrate (i.e., ATP) in a catalytically competent position, thereby allowing the active-site located Asp_{299} to accept a proton from the 5'-group of the ribose of Ado resulting in direct nucleophilic attack on the terminal phosphate of ATP by an in-line S_N2 mechanism. Results further suggested that during the whole process, Arg_{131} acts as the bidentate electrophile. First, it stabilises the resulting quinquivalent transition state by interacting with two negatively charged oxygen groups of the terminal phosphate of ATP and second, Arg_{131} possibly helps in increasing the electrophilicity of the γ-phosphorus atom by withdrawing the negative charge of the oxygen atoms.

Product-Mediated Enzyme Regulation

LdAdK, similar to AdK from most sources, is known to be inhibited by AMP and ADP, raising the possibility of product-mediated regulation.[59,61,79] It is well known that AMP is a competitive inhibitor of the enzyme with respect to Ado and noncompetitive with respect to ATP. In contrast, ADP behaved as a noncompetitive inhibitor with respect to both Ado and ATP, with inhibition by ADP becoming uncompetitive at higher concentrations of ATP.[59] However, until recently, it was not known as to whether the same amino acids were involved in binding both Ado and AMP. Moreover, very little information with regard to ADP binding site were available. However, development of the Asp$_{16}$ mutant, defective in Ado binding, permitted investigations on the mechanism of AMP inhibition. In these studies it has been shown that although Ado and AMP occupy a nearly overlapping position resulting in apparent competition between the two, their mode of interaction with the enzyme are not exactly identical. Analysis suggested that Arg$_{131}$, which has been identified as the key residue involved in the phosphotransfer mechanism, plays an additional role in AMP binding, thereby acting as an effector for product-mediated enzyme regulation. This dual role of Arg$_{131}$ (both in catalysis as well as in regulation) has been further supported with the help of the AMP-docked structure of LdAdK.[65]

Apart from the AMP-mediated regulation, ADP also appears to regulate the activity of the enzyme. Although evidence in favour of this notion is still scanty and will require extensive investigations, the available in vitro results suggest that unlike AMP, the ADP-mediated regulatory mechanism involves the simultaneous participation of another chaperonic protein viz. cyclophilin (CyP). It has recently been demonstrated that LdAdK, which has an inherent tendency to form inactive soluble oligomers, could be disaggregated by a cyclophilin from *L. donovani* (LdCyP) in an isomerase-independent fashion, resulting in reactivation.[80] The reactivation of LdAdK could be achieved in vitro with a stoichiometric amount of LdCyP and under simulated in vivo condition.[81] While investigating the mechanism, it was discovered that ADP, generated during the AdK reaction, facilitates formation of these AdK aggregates, leading to its inactivation. Detailed analysis of the mechanism of reactivation suggested that LdCyP-induced reactivation occurs due to conformational reorientation of AdK in a manner that decreases the affinity of the enzyme for ADP resulting in disaggregation of the inactive oligomers to active monomers.[82] A mechanism of ADP-mediated regulation of LdAdK has also been proposed (Fig. 6).

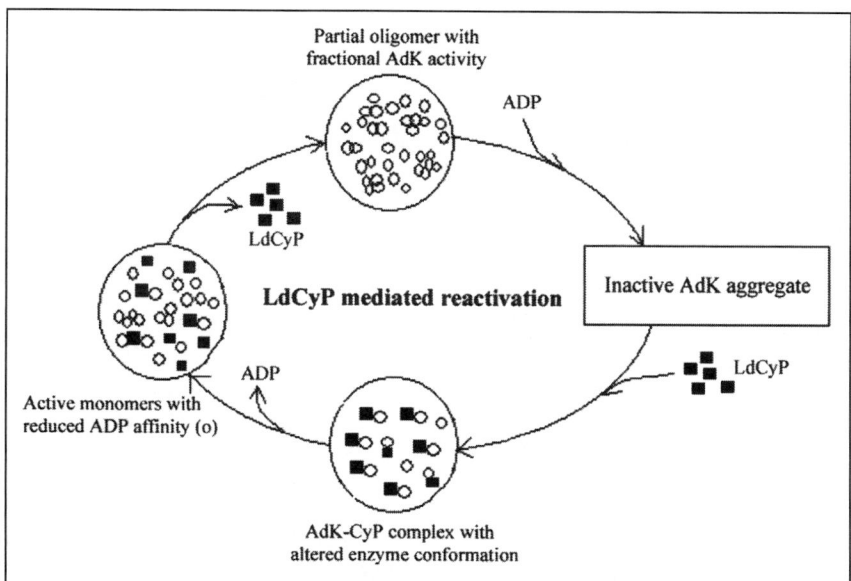

Figure 6. Schematic representation of cyclophilin mediated reactivation of adenosine kinase. Reprinted with permission from: Sen et al, Biochemistry 2006; 45:263-271; ©2006 American Chemical Society.[82]

Since, nucleotide-induced aggregation-disaggregation of enzymes forms the basis of enzyme regulation in many cases, the likelihood of this mechanism operating in *L. donovani* cannot be ruled out.[82]

Based on these observations, it has been possible to arrive at a point from where the "catalytic movie" of LdAdK during the progression of reaction can be speculated (Fig. 7). The self-explanatory cartoon pinpoints the likely conformational change that possibly occurs during the overall process. The ADP-induced aggregation of the enzyme, which may form a basis for enzyme regulation, has however been excluded from this proposed mechanism.

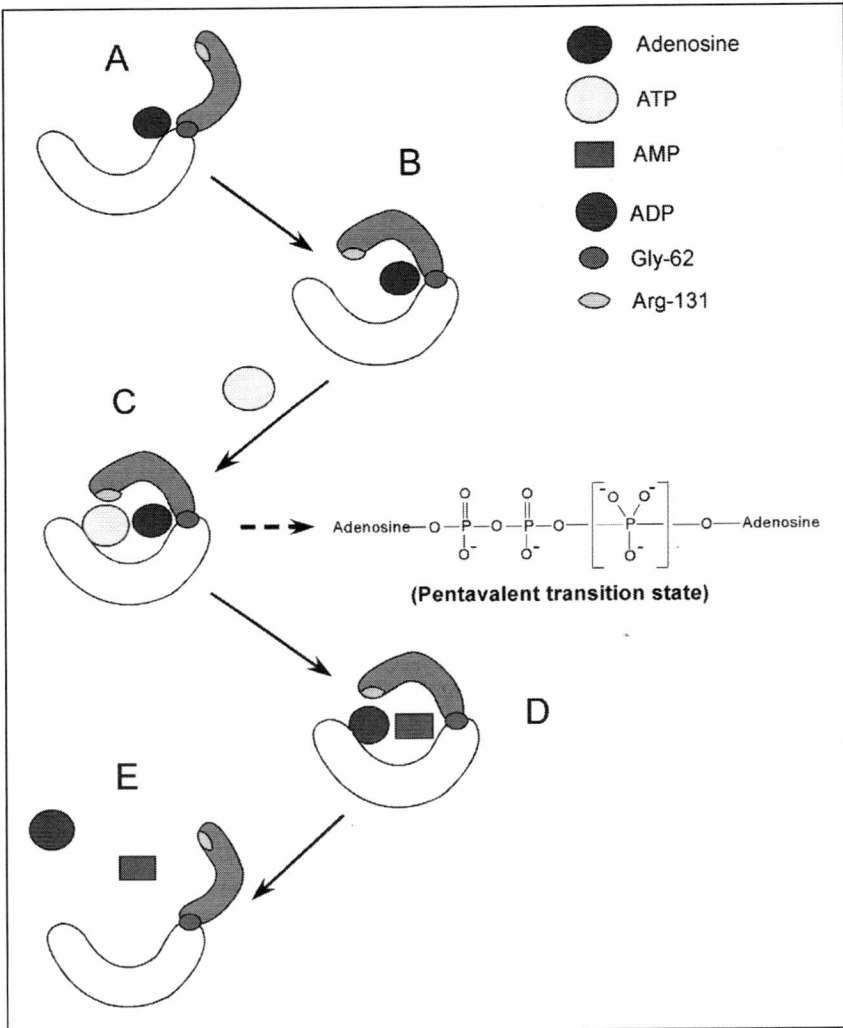

Figure 7. Cartoon representation of the proposed LdAdK catalysed reaction. A) Ado binds to the open active site of the enzyme followed by, B) a rotation of the small domain around the di-glycine hinge causing domain closure and recruitment of Arg-131 to the active site; C) Binding of ATP to the closed active site resulting in phosphate transfer via a pentavalent transition state; D) Formation of the products and, E) Subsequent product release.

Conclusions and Perspectives

Extensive research over the last several years has identified several unique metabolic pathways obligatory to survival and multiplication of parasites. Of these prospective pathways, this chapter has tried to articulate the importance of the pathway responsible for the salvage of purine nucleotides of the purine auxotrophic parasitic protozoa, with special reference to AdK of *Leishmania*. Our interest in this pathway stemmed from the observation that LdAdK, being the gateway enzyme for adenosine nucleotide assimilation in *Leishmania*, shows stage-specific activity profile during morphogenic transformation of the parasite and the possibility of synthesizing subversive nucleoside analogs capable of selectively inhibiting the AdK-mediated phosphorylation reaction in *Leishmania* exists. Our aim has been to address two specific questions: (i) what structural features of this parasitic enzyme are potentially important for the process of phosphate transfer and substrate binding and (ii) how do these features relate to the transition state of catalysis and the overall reaction mechanism? To this end, homology modeling of LdAdK has allowed visualization of the active site of the enzyme and analyse the results of the mutagenesis experiments. Although the actual crystal structure of the protein is a must, the model would be useful in identifying additional sites for mutagenesis and conceptualising the results until the structure of LdAdK is determined experimentally. An additional point of significance of these findings is that, by understanding the structural requirements of product binding, one can certainly conceive of strategies for designing inhibitors capable of interacting with the product binding sites. Therefore, studies directed towards exploiting LdAdK as the target for designing structure-based inhibitor or other enzymes of the purine salvage pathway might prove rewarding.

Acknowledgement

Sincere efforts have been made to acknowledge the contributions of a large number of workers active in this field. Despite this, there are bound to be some inadvertent omissions for which we apologize. Rupak Datta and Banibrata Sen were each supported from the individual fellowship grant from the Council of Scientific and Industrial Research, New Delhi.

References

1. Chang KP, Fong D, Bray RS, eds. Biology of Leishmania and leishmaniasis in "Leishmaniasis". Volume 1. Amsterdam: Elsevier Science Publishers BV, 1985:1-30.
2. Barrett MP, Mottram JC, Coombs GH. Recent advances in identifying and validating drug targets in trypanosomes and leishmanias. Trends Microbiol 1999; 7(2):82-88.
3. Killick-Kendrick R, Molyneux DH, Hommel M et al. Leishmania in phlebotomid sandflies. V. The nature and significance of infections of the pylorus and ileum of the sandfly by leishmaniae of the braziliensis complex. Proc R Soc Lond B Biol Sci 1977; 198(1131):191-199.
4. Chang KP. Leishmania donovani: Promastigote—macrophage surface interactions in vitro. Exp Parasitol 1979; 48(2):175-189.
5. el Kouni MH. Potential chemotherapeutic targets in the purine metabolism of parasites. Pharmacol Ther 2003; 99(3):283-309.
6. Pratt DM, David JR. Monoclonal antibodies recognizing determinants specific for the promastigote state of Leishmania mexicana. Mol Biochem Parasitol 1982; 6(5):317-327.
7. Jaffe CL, Bennett E, Grimaldi Jr G et al. Production and characterization of species-specific monoclonal antibodies against Leishmania donovani for immunodiagnosis. J Immunol 1984; 133(1):440-447.
8. Looker DL, Berens RL, Marr JJ. Purine metabolism in Leishmania donovani amastigotes and promastigotes. Mol Biochem Parasitol 1983; 9(1):15-28.
9. Dwyer DM, Langreth SG, Dwyer NK. Evidence for a polysaccharide surface coat in the developmental stages of Leishmania donovani: A fine structure-cytochemical study. Z Parasitenkd 1974; 43(4):227-249.
10. Janovy Jr J. Respiratory changes accompanying Leishmania to leptomonad transformation in Leishmania donovani. Exp Parasitol 1967; 20(1):51-55.
11. Krassner SM, Morrow CD, Flory B. Inhibition of Leishmania donovani amastigote-to-promastigote transformation by infected hamster spleen lymphocyte lysates. J Protozool 1980; 27(1):87-92.

12. Konigk E, Putfarken B. Stage-specific differences of a perhaps signal-transferring system in Leishmania donovani. Tropenmed Parasitol 1980; 31(4):421-424.

13. Peterson DS, Milhous WK, Wellems TE. Molecular basis of differential resistance to cycloguanil and pyrimethamine in Plasmodium falciparum malaria. Proc Natl Acad Sci USA 1990; 87(8):3018-3022.

14. Ring CS, Sun E, McKerrow JH et al. Structure-based inhibitor design by using protein models for the development of antiparasitic agents. Proc Natl Acad Sci USA 1993; 90(8):3583-3587.

15. Foote SJ, Galatis D, Cowman AF. Amino acids in the dihydrofolate reductase-thymidylate synthase gene of Plasmodium falciparum involved in cycloguanil resistance differ from those involved in pyrimethamine resistance. Proc Natl Acad Sci USA 1990; 87(8):3014-3017.

16. Trager W. Recent progress in some aspects of the physiology of parasitic protozoa. J Parasitol 1970; 56(4):627-633.

17. Walsh CJ, Sherman IW. Purine and pyrimidine synthesis by the avian malaria parasite, Plasmodium lophurae. J Protozool 1968; 15(4):763-770.

18. Bone GJ, Steinert M. Isotopes incorporated in the nucleic acids of Trypanosoma mega. Nature 1956; 178(4528):308-309.

19. Hammond DJ, Gutteridge WE. Purine and pyrimidine metabolism in the Trypanosomatidae. Mol Biochem Parasitol 1984; 13(3):243-261.

20. Vasudevan G, Carter NS, Drew ME et al. Cloning of Leishmania nucleoside transporter genes by rescue of a transport-deficient mutant. Proc Natl Acad Sci USA 1998; 95(17):9873-9878.

21. Carter NS, Drew ME, Sanchez M et al. Cloning of a novel inosine-guanosine transporter gene from Leishmania donovani by functional rescue of a transport-deficient mutant. J Biol Chem 2000; 275(27):20935-20941.

22. Landfear SM. Molecular genetics of nucleoside transporters in Leishmania and African trypanosomes. Biochem Pharmacol 2001; 62(2):149-155.

23. Gottlieb M, Dwyer DM. Protozoan parasite of humans: Surface membrane with externally disposed acid phosphatase. Science 1981; 212(4497):939-941.

24. Gottlieb M, Dwyer DM. Leishmania donovani: Surface membrane acid phosphatase activity of promastigotes. Exp Parasitol 1981; 52(1):117-128.

25. Bates PA, Dwyer DM. Biosynthesis and secretion of acid phosphatase by Leishmania donovani promastigotes. Mol Biochem Parasitol 1987; 26(3):289-296.

26. Debrabant A, Bastien P, Dwyer DM. A unique surface membrane anchored purine-salvage enzyme is conserved among a group of primitive eukaryotic human pathogens. Mol Cell Biochem 2001; 220(1-2):109-116.

27. Glew RH, Saha AK, Das S et al. Biochemistry of the Leishmania species. Microbiol Rev 1988; 52(4):412-432.

28. Jardim A, Bergeson SE, Shih S et al. Xanthine phosphoribosyltransferase from Leishmania donovani. Molecular cloning, biochemical characterization, and genetic analysis. J Biol Chem 1999; 274(48):34403-34410.

29. LaFon SW, Nelson DJ, Berens RL et al. Inosine analogs. Their metabolism in mouse L cells and in Leishmania donovani. J Biol Chem 1985; 260(17):9660-9665.

30. Hassan HF, Coombs GH. Leishmania mexicana: Purine-metabolizing enzymes of amastigotes and promastigotes. Exp Parasitol 1985; 59(2):139-150.

31. Ghosh M, Mukherjee T. Stage-specific development of a novel adenosine transporter in Leishmania donovani amastigotes. Mol Biochem Parasitol 2000; 108(1):93-99.

32. Hwang HY, Ullman B. Genetic analysis of purine metabolism in Leishmania donovani. J Biol Chem 1997; 272(31):19488-19496.

33. Tuttle JV, Krenitsky TA. Purine phosphoribosyltransferases from Leishmania donovani. J Biol Chem 1980; 255(3):909-916.

34. Marr JJ. Pyrazolopyrimidine metabolism in Leishmania and trypanosomes: Significant differences between host and parasite. J Cell Biochem 1983; 22(3):187-196.

35. Fish WR, Marr JJ, Berens RL et al. Inosine analogs as chemotherapeutic agents for African trypanosomes: Metabolism in trypanosomes and efficacy in tissue culture. Antimicrob Agents Chemother 1985; 27(1):33-36.

36. Martinez S, Marr JJ. Allopurinol in the treatment of American cutaneous leishmaniasis. N Engl J Med 1992; 326(11):741-744.

37. Kager PA, Rees PH, Wellde BT et al. Allopurinol in the treatment of visceral leishmaniasis. Trans R Soc Trop Med Hyg 1981; 75(4):556-559.

38. Somoza JR, Chin MS, Focia PJ et al. Crystal structure of the hypoxanthine-guanine-xanthine phosphoribosyltransferase from the protozoan parasite Tritrichomonas foetus. Biochemistry 1996; 35(22):7032-7040.

39. Schumacher MA, Carter D, Ross DS et al. Crystal structures of Toxoplasma gondii HGXPRTase reveal the catalytic role of a long flexible loop. Nat Struct Biol 1996; 3(10):881-887.

40. Focia PJ, Craig IIIrd SP, Nieves-Alicea R et al. A 1.4 A crystal structure for the hypoxanthine phosphoribosyltransferase of Trypanosoma cruzi. Biochemistry 1998; 37(43):15066-15075.

41. Freymann DM, Wenck MA, Engel JC et al. Efficient identification of inhibitors targeting the closed active site conformation of the HPRT from Trypanosoma cruzi. Chem Biol 2000; 7(12):957-968.

42. Aronov AM, Munagala NR, Ortiz De Montellano PR et al. Rational design of selective submicromolar inhibitors of Tritrichomonas foetus hypoxanthine-guanine-xanthine phosphoribosyltransferase. Biochemistry 2000; 39(16):4684-4691.

43. Somoza JR, Skillman Jr AG, Munagala NR et al. Rational design of novel antimicrobials: Blocking purine salvage in a parasitic protozoan. Biochemistry 1998; 37(16):5344-5348.

44. Iltzsch MH, Uber SS, Tankersley KO et al. Structure-activity relationship for the binding of nucleoside ligands to adenosine kinase from Toxoplasma gondii. Biochem Pharmacol 1995; 49(10):1501-1512.

45. Cohen SS, Plunkett W. The utilization of nucleotides by animal cells. Ann NY Acad Sci 1975; 255(751106-751230-2):269-286.

46. Krug EC, Marr JJ, Berens RL. Purine metabolism in Toxoplasma gondii. J Biol Chem 1989; 264(18):10601-10607.

47. Pfefferkorn ER, Pfefferkorn LC. Arabinosyl nucleosides inhibit Toxoplasma gondii and allow the selection of resistant mutants. J Parasitol 1976; 62(6):993-999.

48. Schwartzman JD, Pfefferkorn ER. Toxoplasma gondii: Purine synthesis and salvage in mutant host cells and parasites. Exp Parasitol 1982; 53(1):77-86.

49. Pfefferkorn ER, Pfefferkorn LC. The biochemical basis for resistance to adenine arabinoside in a mutant of Toxoplasma gondii. J Parasitol 1978; 64(3):486-492.

50. Iovannisci DM, Ullman B. Characterization of a mutant Leishmania donovani deficient in adenosine kinase activity. Mol Biochem Parasitol 1984; 12(2):139-151.

51. Datta AK, Bhaumik D, Chatterjee R. Isolation and characterization of adenosine kinase from Leishmania donovani. J Biol Chem 1987; 262(12):5515-5521.

52. Chaudhary K, Darling JA, Fohl LM et al. Purine salvage pathways in the apicomplexan parasite Toxoplasma gondii. J Biol Chem 2004; 279(30):31221-31227.

53. el Kouni MH, Guarcello V, Al Safarjalani ON et al. Metabolism and selective toxicity of 6-nitrobenzylthioinosine in Toxoplasma gondii. Antimicrob Agents Chemother 1999; 43(10):2437-2443.

54. Yadav V, Chu CK, Rais RH et al. Synthesis, biological activity and molecular modeling of 6-benzylthioinosine analogues as subversive substrates of Toxoplasma gondii adenosine kinase. J Med Chem 2004; 47(8):1987-1996.

55. Rais RH, Al Safarjalani ON, Yadav V et al. 6-Benzylthioinosine analogues as subversive substrate of Toxoplasma gondii adenosine kinase: Activities and selective toxicities. Biochem Pharmacol 2005; 69(10):1409-1419.

56. Mathews II, Erion MD, Ealick SE. Structure of human adenosine kinase at 1.5 A resolution. Biochemistry 1998; 37(45):15607-15620.

57. Schumacher MA, Scott DM, Mathews II et al. Crystal structures of Toxoplasma gondii adenosine kinase reveal a novel catalytic mechanism and prodrug binding. J Mol Biol 2000; 298(5):875-893.

58. Bhaumik D, Datta AK. Immunochemical and catalytic characteristics of adenosine kinase from Leishmania donovani. J Biol Chem 1989; 264(8):4356-4361.

59. Bhaumik D, Datta AK. Reaction kinetics and inhibition of adenosine kinase from Leishmania donovani. Mol Biochem Parasitol 1988; 28(3):181-187.

60. Sinha KM, Ghosh M, Das I et al. Molecular cloning and expression of adenosine kinase from Leishmania donovani: Identification of unconventional P-loop motif. Biochem J 1999; 339(Pt 3):667-673.

61. Palella TD, Andres CM, Fox IH. Human placental adenosine kinase. Kinetic mechanism and inhibition. J Biol Chem 1980; 255(11):5264-5269.

62. Singh B, Hao W, Wu Z et al. Cloning and characterization of cDNA for adenosine kinase from mammalian (Chinese hamster, mouse, human and rat) species. High frequency mutants of Chinese hamster ovary cells involve structural alterations in the gene. Eur J Biochem 1996; 241(2):564-571.

63. Wu LF, Reizer A, Reizer J et al. Nucleotide sequence of the Rhodobacter capsulatus fruK gene, which encodes fructose-1-phosphate kinase: Evidence for a kinase superfamily including both phosphofructokinases of Escherichia coli. J Bacteriol 1991; 173(10):3117-3127.

64. Bork P, Sander C, Valencia A. Convergent evolution of similar enzymatic function on different protein folds: The hexokinase, ribokinase, and galactokinase families of sugar kinases. Protein Sci 1993; 2(1):31-40.

65. Datta R, Das I, Sen B et al. Homology-model-guided site-specific mutagenesis reveals the mechanisms of substrate binding and product-regulation of adenosine kinase from Leishmania donovani. Biochem J 2006; 394(Pt 1):35-42.

66. Sigrell JA, Cameron AD, Jones TA et al. Structure of Escherichia coli ribokinase in complex with ribose and dinucleotide determined to 1.8 A resolution: Insights into a new family of kinase structures. Structure 1998; 6(2):183-193.

67. Saraste M, Sibbald PR, Wittinghofer A. The P-loop—a common motif in ATP- and GTP-binding proteins. Trends Biochem Sci 1990; 15(11):430-434.

68. Muller CW, Schulz GE. Structure of the complex between adenylate kinase from Escherichia coli and the inhibitor Ap5A refined at 1.9 A resolution. A model for a catalytic transition state. J Mol Biol 1992; 224(1):159-177.

69. Matte A, Tari LW, Delbaere LT. How do kinases transfer phosphoryl groups? Structure 1998; 6(4):413-419.

70. Story RM, Steitz TA. Structure of the recA protein-ADP complex. Nature 1992; 355(6358):374-376.

71. Berchtold H, Reshetnikova L, Reiser CO et al. Crystal structure of active elongation factor Tu reveals major domain rearrangements. Nature 1993; 365(6442):126-132.

72. Muegge I, Schweins T, Langen R et al. Electrostatic control of GTP and GDP binding in the oncoprotein p21ras. Structure 1996; 4(4):475-489.

73. Smith CM, Radzio-Andzelm E, Madhusudan et al. The catalytic subunit of cAMP-dependent protein kinase: Prototype for an extended network of communication. Prog Biophys Mol Biol 1999; 71(3-4):313-341.

74. Van der Ploeg LH. Discontinuous transcription and splicing in trypanosomes. Cell 1986; 47(4):479-480.

75. Darling JA, Sullivan Jr WJ, Carter D et al. Recombinant expression, purification, and characterization of Toxoplasma gondii adenosine kinase. Mol Biochem Parasitol 1999; 103(1):15-23.

76. Carret C, Delbecq S, Labesse G et al. Characterization and molecular cloning of an adenosine kinase from Babesia canis rossi. Eur J Biochem 1999; 265(3):1015-1021.

77. Spychala J, Datta NS, Takabayashi K et al. Cloning of human adenosine kinase cDNA: Sequence similarity to microbial ribokinases and fructokinases. Proc Natl Acad Sci USA 1996; 93(3):1232-1237.

78. Datta R, Das I, Sen B et al. Mutational analysis of the active-site residues crucial for catalytic activity of adenosine kinase from Leishmania donovani. Biochem J 2005; 387(Pt 3):591-600.

79. Hawkins CF, Bagnara AS. Adenosine kinase from human erythrocytes: Kinetic studies and characterization of adenosine binding sites. Biochemistry 1987; 26(7):1982-1987.

80. Chakraborty A, Das I, Datta R et al. A single-domain cyclophilin from Leishmania donovani reactivates soluble aggregates of adenosine kinase by isomerase-independent chaperone function. J Biol Chem 2002; 277(49):47451-47460.

81. Chakraborty A, Sen B, Datta R et al. Isomerase-independent chaperone function of cyclophilin ensures aggregation prevention of adenosine kinase both in vitro and under in vivo conditions. Biochemistry 2004; 43(37):11862-11872.

82. Sen B, Chakraborty A, Datta R et al. Reversal of ADP-Mediated Aggregation of Adenosine Kinase by Cyclophilin Leads to Its Reactivation. Biochemistry 2006; 45(1):263-271.

CHAPTER 11

Searching the Tritryp Genomes for Drug Targets

Peter J. Myler*

Abstract

The recent publication of the complete genome sequences of *Leishmania major*, *Trypanosoma brucei* and *Trypanosoma cruzi* revealed that each genome contains 8300-12,000 protein-coding genes, of which ~6500 are common to all three genomes, and ushers in a new, post-genomic, era for trypanosomatid drug discovery. This vast amount of new information makes possible more comprehensive and accurate target identification using several new computational approaches, including identification of metabolic "choke-points", searching the parasite proteomes for orthologues of known drug targets, and identification of parasite proteins likely to interact with known drugs and drug-like small molecules. In this chapter, we describe several databases (such as GENEDB, BRENDA, KEGG, METACYC, the THERA-PEUTIC TARGET DATABASE, and CHEMBANK) and algorithms (including PATHOLOGIC, PATHWAY HUNTER TOOL, AND AUTODOCK) which have been developed to facilitate the bioinformatic analyses underlying these approaches. While target identification is only the first step in the drug development pipeline, these new approaches give rise to renewed optimism for the discovery of new drugs to combat the devastating diseases caused by these parasites.

Traditionally, drug discovery in the trypanosomatids (and other organisms) has proceeded from two different starting points: screening large numbers of existing compounds for activity against whole parasites or more focused screening of compounds for activity against defined molecular targets. Most existing anti-trypanosomatids drugs were developed using the former approach, although the latter has gained much attention in the last twenty years under the rubric of "rational drug design". Until recently, one of the major bottlenecks in anti-trypanosomatid drug development has been our ability to identify good targets, since only a very small percentage of the total number of trypanosomatid genes were known. That has now changed forever, with the recent (July, 2005) publication of the "Tritryp" (*Trypanosoma brucei*, *Trypanosoma cruzi* and *Leishmania major*) genome sequences.[1-4] This vast amount of information now makes possible several new approaches for target identification and ushers in a post-genomic era for trypanosomatid drug discovery.

Tritryp Genome Content

According to the latest data released at GeneDB (http://www.genedb.org), the haploid genomes of *T. brucei*, *T. cruzi* and *L. major* encode 9878, ~12000, and 8373 likely protein-coding genes (and pseudogenes), respectively (see Table 1). The gene densities of the two trypanosome genomes are quite similar (300-400 genes/Mb) and somewhat higher than that of *L. major* (250 gene/Mb). The average coding sequence (CDS) is slightly larger in *Leishmania*, as a result

*Peter J. Myler—Seattle Biomedical Research Institute, 307 Westlake Ave N., Suite 500, Seattle, Washington 98109-5219, USA. Email: peter.myler@sbri.org

Drug Targets in Kinetoplastid Parasites, edited by Hemanta K. Majumder.
©2008 Landes Bioscience and Springer Science+Business Media.

Table 1. Tritryp genome statistics

	T. brucei	T. cruzi	L. major
Protein-coding genes	8599[a]	~10,000[b]	8302
Pseudogenes	1279	~2000[c]	71
Average CDS length (bp)	1,592	1,513	1,901
Average inter-CDS size	1,279	1,024	2,045
Gene density (gene/Mb)	317	385	252
Function known	5%	n.d.[d]	4%
Function inferred	38%	43%	28%
Hypothetical, conserved	51%	48%	56%
Hypothetical, species-specific	6%	9%	8%
Orthologues in all Tritryps	73%	54%	80%
Tb+Tc only	5%	4%	-
Tb+Lm only	1%	-	1%
Tc+Lm only	-	4%	6%
Species-specific	21%	38%	13%

a) Excludes 612 genes annotated as hypothetical protein, unlikely. b) Total number in both haplotypes is 18,980. c) Number in both haplotypes is 3,590. d) Not determined.

of small sequence insertions relative to the trypanosomes, but the lower gene density in *Leishmania* is mostly explained by its larger inter-CDS regions. Each species contains a number of gene families of varying size. Predicted functions have been ascribed to ~40% of the protein-coding genes, but this has been confirmed experimentally for only ~5% of the proteins. Most of the remaining genes encode conserved hypothetical proteins, of which slightly more than half are found only in trypanosomatids. Interestingly, ~2-3% of the Tritryp proteins are related to those found in prokaryotes but not other eukaryotes. At least some of these appear to have arisen from horizontal gene transfer, and may represent excellent candidates for drug targets. The Tritryp genomes display a remarkable degree of synteny, with ~75% of the genes in *L. major* having orthologues in both other species and >90% of these occurring in the same genomic context (see Table 1). The proteins within this Tritryp "core" proteome exhibit an average 57% identity between *T. brucei* and *T. cruzi*, and 44% identity between *L. major* and the two other trypanosomes, reflecting the expected phylogenetic relationships.[5,6] Interestingly, substantially fewer orthologues are shared only between *L. major* and *T. brucei* than between *L. major* and *T. cruzi*, perhaps reflecting the common intracellular environment of their mammalian stages.

However, all three genomes contain a significant number of species-specific genes, which account for ~21% and 38% of the protein-coding genes in *T. brucei* and *T. cruzi*, respectively, but only ~13% of the *L. major* genes. These species-specific genes (and pseudogenes) mostly encode large families of surface proteins, exemplified by the variant surface glycoproteins (VSGs) and Procyclic Acidic Repetitive Proteins (EP/PARP/procyclin) of *T. brucei*; the trans-sialidases, dispersed gene family protein 1 (DGF-1), mucins, and mucin-associated surface proteins (MASPs) of *T. cruzi*; and the amastins and promastigote surface antigens (PSA-2) of *L. major*. In addition to these species-specific genes, all three species demonstrate differential paralogous gene expansion or contraction, with the ESAG4 adenylate/guanylate cyclases and leucine-rich repeat proteins being over-represented in *T. brucei*; GP63 surface proteases and recombination hot spot (RHS) proteins in *T. cruzi*; and mitochondrial carrier protein, ATP-Binding Cassette (ABC) transporters, and Heat Shock Protein (HSP) 90 gene families in *L. major*. Many of these species-specific genes or paralogous expansions occur in telomeric and sub-telomeric gene clusters, possibly reflecting similar strategies used for immune evasion.

Transcription and RNA processing in the trypanosomatids is quite different from that in other eukaryotes,[7] with unique or unusual processes such as large polycistronic gene clusters,[8-10] RNA polymerase I-mediated transcription of some protein-coding genes,[11,12] and trans-splicing.[13] While annotation of the Tritryp genomes uncovered most of the expected RNAP polymerase subunits, there was a dearth of transcription factors normally involved in regulation of transcription initiation by other eukaryotes.[3] However, recent experiments have identified several highly divergent transcription factors in *T. brucei*,[14-17] suggesting that Tritryp transcription initiation may represent an ancestral, less sequence-specific, mechanism mostly replaced in other eukaryotes by the archetypal TATA-containing promoters. Conversely, the paucity of Tritryp genes encoding transcriptional regulators is offset by an abundance of proteins with RNA binding motifs,[18] consistent with their reliance on post-transcriptional models of gene regulation.[19]

DNA replication in trypanosomatids also appears to differ significantly from that in higher eukaryotes, with only one of the six subunits typically found in the eukaryotic replication origin complex being identified.[2] There are also substantial differences in the mitochondrial replication machinery, since the complexity of the kinetoplast DNA (the trypanosomatid equivalent of a mitochondrial genome) structure dictates an unusual replication mechanism.[20]

Bioinformatic analyses of the Tritryp genomes suggests that they lack several classes of signaling molecules found in other eukaryotes, including serpentine receptors, heterotrimeric G proteins, most classes of catalytic receptors, SH2 and SH3 interaction domains, and regulatory transcription factors, but that they do possess a large and complex set of protein kinases and protein phosphatases.[2,21] However, the distribution of protein kinase classes differs from that in other organisms; with no tyrosine kinases (other than dual specificity kinases), receptor kinases or TKL and RGC group kinases. Since the trypanosomatids have complicated life cycles in different hosts, it is likely that these kinases play important roles in regulating their response to changes in these different environments.

Computational Approaches for Drug Target Selection

The experience gained by the pharmaceutical industry during the last few decades of drug development has lead to the postulation of a number of selection criteria for successful drug target identification.[22] In the context of the trypanosomatids, these criteria include selectivity (i.e., the parasite target is absent from, or substantially different in, the host); "druggability" (the target structure has a small molecule-binding pocket); suitable biochemical properties (the target has a low turnover rate and/or catalyzes a rate-limiting step within a pathway); validation (the target is essential for growth and/or survival in the mammalian stage of the parasite lifecycle); "assayability" (specific, inexpensive and high-throughput screens are available using in vitro expressed target); and low potential for development of drug resistance (absence of different isoforms or alleles and/or biochemical "bypass" reactions). With these criteria in mind, several bioinformatic approaches have been proposed, which take advantage of the availability of the complete genome sequences described above to accelerate progress in developing effective clinical interventions for the important diseases caused by these parasites.

Analysis of the Tritryp genomes has provided a comprehensive view of the parasites' metabolic potential by identifying numerous common and species-specific metabolic and transport processes. Manual examination of metabolic maps identified a number of pathways that appear to be especially amenable to potential chemotherapeutic intervention; including glycolysis, the electron transport chain, the urea cycle, the glyoxylase pathway and associated trypanothione metabolism, glycosylphosphatidylinositol (GPI) anchor biosynthesis, fatty acid biosynthesis, as well as the ergosterol and isoprenoid biosynthetic pathway.[1] Since the particulars of target identification and drug development for each of these pathways (and others) have been described in detail in several of the accompanying chapters and elsewhere,[23-26] they will not be further explored here. Instead, several different computational attempts to catalogue metabolic pathways and identify "choke-points" will be described.

Databases of Tritryp Metabolism

BRENDA (BRaunschweig ENzyme DAtabase) is a comprehensive collection of enzyme and metabolic information (http://www.brenda.uni-koeln.de), including Enzyme Commission (EC) classification and nomenclature, reaction and specificity, function and structure, isolation and stability, as well as links to primary literature references. The database is now based on a controlled vocabulary and ontology for some information fields, and search tools include EC and taxonomy-tree browsers, a chemical substructure search engine for ligand structure, and a thesaurus for ligand names. BRENDA contains more than 100,000 enzymes representing 4060 different EC numbers from about ~10000 different organisms. There are currently (as of September, 2006) 842 entries for *T. brucei*, 751 for *T. cruzi* and 607 for *L. major*.

KEGG (Kyoto Encyclopedia of Genes and Genomes) is a suite of databases and associated software, designed to integrate current knowledge of genes and proteins (GENES database), chemical compounds and reactions (LIGAND), metabolic, regulatory and interaction networks (PATHWAY), and ontologies (BRITE). Biological systems are represented in KEGG by nested graphs, which are used for pathway reconstruction and functional inference, and line graphs, which form the basis for integrating genome and chemical information with the networks. The BRITE database provides the pathway reconstruction through a series of functional hierarchies and represents the logical foundation for the KEGG project. KEGG maintains a gene catalogue of sequenced genomes and maps them onto 301 manually drawn and curated reference pathways.[27-31] Currently, there are 83, 90, and 89 entries in the PATHWAY database for *T. brucei*, *T. cruzi* and *L. major*, respectively, mostly describing metabolic pathways.

The BIOCYC collection of Pathway/Genome Databases (PGDBs) provides electronic reference sources on the pathways and genomes of more than 200 different organisms (http://biocyc.org). The databases within the BIOCYC collection are organized into tiers according to the amount of manual review and updating they have received. Tier 1 PGDBs are created through intensive manual efforts, and receive continuous updating. EcoCyc, which describes *Escherichia coli* K-12, is the only organism-specific Tier 1 database. Tier 2 PGDBs are computationally generated using PATHOLOGIC software,[32,33] and have undergone moderate amounts of review and updating. There are currently 12 databases in Tier 2, including HUMANCYC and PLASMOCYC (which describes the malaria parasite, *Plasmodium falciparum*). Tier 3 databases are computationally generated by the PATHOLOGIC program, and have undergone no review and updating.[34] There are 191 PGDBs in Tier 3, representing mostly bacterial genomes. The individual BIOCYC web-sites can be used to visualize single or multiple metabolic pathways, including a complete metabolic map of the organism. An OMICS VIEWER can be used to analyze gene expression, proteomics, or metabolomics data to produce animated views of time-course gene-expression experiments. There are currently no BIOCYC PGDBs for any of the trypanosomatid genomes, although it should be relatively straightforward to generate Tier 3 databases using the PATHOLOGIC software.[32] Other programs are also available for genome-scale reconstruction of metabolic networks.[35-38] However, since this process is largely dependent on sequence-based homology searches to identify the enzymes and the Tritryp genomes are quite divergent from other eukaryotes, considerable manual curation will probably be necessary to obtain truly accurate representations of the metabolic networks in these organisms.

While most of the individual PGDBs within BIOCYC represent species-specific databases, METACYC (http://metacyc.org) is a collection of metabolic pathways and enzymes from more than 240 organisms (mostly bacteria and plants). The goal of METACYC is to represent every experimentally elucidated metabolic pathway, reaction, and chemical compound, as well as the genes encoding the enzymes that catalyze the reactions involved.[39] As well as being used as a reference source to look up individual facts, METACYC facilitates computational studies of the metabolism, such as design of novel biochemical pathways for biotechnology, studies of evolution of metabolic pathways, and simulation of metabolic pathways. Additionally, desktop software is available for comparing the overall metabolic maps, specific pathways and genomic maps of two organisms.

Identification of Metabolic "Choke-Points"

Careful manual examination of a metabolic pathway can identify metabolic "choke-points", i.e., the enzyme(s) which is (are) uniquely necessary to produce a critical metabolite. Obviously, choke-points in pathways that result in metabolites critical for parasite survival would make excellent potential targets for development of novel anti-trypanosomatid drugs. The PATHWAY HUNTER TOOL (http://www.pht.uni-koeln.de) uses an extended form of graph theory (in which enzymes are represented by edges between nodes representing metabolites) to identify choke-points and rank them according to their "load".[40,41] Load is defined as the ratio of the number of shortest paths through the enzyme and nearest neighbors attached to it, compared to the average values for these properties in the entire network. Comparison of pathogen (trypanosomatid) and host (human) metabolic networks could be used to identify highly ranked choke-points that are unique to the parasite or are ranked much lower in the host.

Another computational approach for identification of metabolic enzymes as drug targets involves the concept of minimal cut sets, which are defined as the minimal set of reaction in a network whose inactivation will definitively lead to a failure in a particular network function.[42] Screening parasite metabolic networks for all possible minimal cut sets and identification of those which are small (i.e., contain few enzymes) and not present in the host could serve to identify potential drug targets.

The approaches outlined above are designed to identify targets that meet only some of the criteria outlined at the beginning of this section; namely they have suitable biochemical properties, are likely to be essential for the parasite, and are sufficiently different from any host homologue. However, alternative approaches seek to make use of the finding that successful drugs have specific structural and physicochemical properties that allow them to be efficacious, bioavailable, and safe. These properties are exemplified by Lipinski's so-called "rule of five".[43] This has lead to the concept of "druggable" proteins, based on their ability to bind potentially effective drug-like small molecules.[44-46] Thus, it makes sense to search the Tritryps genomes for proteins that are likely to meet these criteria. Two different approaches have been proposed for developing computational solutions to this problem: searching the genome for proteins with similar properties to known drug targets in other organisms (primarily humans) and direct interrogation of the parasite proteins for their likelihood to bind drug-like chemicals.

Searching for Parasite Orthologues of Known Drug Targets

The Therapeutic Target Database (TTD) (http://xin.cz3.nus.edu.sg/group/cjttd/ttd.asp) represents a comprehensive and publicly available attempt to catalogue information about all the currently known protein and nucleic acid targets described in the literature.[46,47] The database also contains information about the drugs and ligands directed at these targets, as well as corresponding disease conditions. This database currently contains 1535 targets and 2107 drugs/ligands, including 19 entries listing potential anti-trypanosomatid use. The most simplistic approach for searching the Tritryp genomes for potential targets similar to these existing targets would be to carry out BLASTP or PSIBLAST searches of the Tritryp protein databases to identify parasite proteins with significant sequence similarity to those in the TTD. The resulting list of parasite proteins would need to be subsequently winnowed down by removing those that are too similar to the human orthologues and/or are similar to proteins involved in more than two pathways in humans, since drugs against these are likely to have deleterious effects on the human host. However, given what we know about the imprecise nature of the relationship between protein sequence and structure, it is likely that this method will have a significant false negative rate (i.e., it will miss many potentially useful targets because they won't have sufficient sequence similarity). Statistical learning methods, such as support vector machines (SVM) and neural networks, have recently enjoyed considerable success for prediction of protein structure and may be useful for identifying targets missed by simple BLAST searching. A SVM method has been used to screen the human and HIV genome for druggable proteins, with a promising degree of accuracy.[46,48,49] Similar methods could be used to screen the Tritryp genomes.

Matching Drug-Like Chemicals to Parasite Proteins

Algorithms such as AutoDock[50] have been used for some time to predict small molecules that will potentially fill protein ligand-binding pockets, as a first step in rational drug design. This process has been reversed to some extent by using docking software with integrated molecular dynamics simulation to predict which drugs are likely to bind (and inhibit) proteases from human coronavirus,[51] cytomegalovirus,[52] and human immunodeficiency virus (HIV).[53] A recent publication describes the use of this method to screen 2500 compounds in the CHEMBANK database (http://chembank.broad.harvard.edu) against 13 proteins from *Plasmodium falciparum* whose structure had been determined by X-ray crystallography.[54] This approach found that the K*i*s predicted for three existing anti-malarial drugs compared well with their known values and that their predicted inhibitory activity ranked in the top 5th percentile of all tested drugs. Another 20 drugs were predicted to have multi-target activity, i.e., they showed high affinity with two or more proteins. Multi-target drugs are attractive because they are less likely to encounter problems with development of drug resistance. It should be possible to screen the Tritryp proteome for multi-target drugs using a similar approach. Obviously, one major constraint is the availability (or lack thereof) of trypanosomatid proteins with known structure. Currently, the protein structure database (PDB) contains 79 nonredundant structures from the genera *Leishmania* or *Trypanosoma*. However, this number has been increasing rapidly over the last few years due to the efforts of the Structural Genomics of Pathogenic Protozoa (SGPP) consortium (http://www.sgpp.org) and is likely to increase further in the near future.

Conclusion

The recent completion of the Tritryp genome sequencing project provides an unprecedented opportunity for development of novel anti-trypanosomatid chemotherapeutic agents. The identification of more than 8000 new protein-coding genes, many of which are shared between the *Leishmania* and *Trypanosoma* genera, vastly expands the potential drug targets available for investigation. In fact, the situation has gone from a relative dearth of useful targets to an embarrassment of riches, with far more potential targets available than can possibly be studied in detail. In this chapter, we have described several different computational approaches that should be useful in reducing this smorgasbord of genes to a manageable number of high-value targets, which will form the basis of detailed biological and pharmacological investigation. Of course, target identification is only the first stages in the lengthy and expensive process of drug development; with steps such as target validation, lead identification and optimization, as well as preclinical pharmacological screening, necessary before a potential drug can enter clinical trials. Nevertheless, these bioinformatic methods hold great promise in being able to identify targets (and potential lead compounds in some cases) which have a higher probability of successful drug development than traditional methods. While only time will reveal the validity of this promise, we hope that this advent of the post-genomics era for trypanosomatid biology heralds a renaissance in the discovery of much needed new drugs for the devastating diseases caused by these parasites.

References

1. Berriman M, Ghedin E, Hertz-Fowler C et al. The genome of the African trypanosome, Trypanosoma brucei. Science 2005; 309:416-422.
2. El-Sayed NMA, Myler PJ, Bartholomeu D et al. The genome sequence of Trypanosoma cruzi, etiological agent of Chagas' disease. Science 2005; 309(5733):409-415.
3. Ivens AC, Peacock CS, Worthey EA et al. The genome of the kinetoplastid parasite, Leishmania major. Science 2005; 309(5733):436-442.
4. El-Sayed NMA, Myler PJ, Blandin G et al. Comparative genomics of trypanosomatid parasitic protozoa. Science 2005; 309(5733):404-409.
5. Haag J, O'hUigin C, Overath P. The molecular phylogeny of trypanosomes: Evidence for an early divergence of the Salivaria. Mol Biochem Parasitol 1998; 91(1):37-49.
6. Stevens JR, Noyes HA, Schofield CJ et al. The molecular evolution of Trypanosomatidae. Adv Parasitol 2001; 48:1-56.

7. Campbell DA, Thomas S, Sturm N. Transcription in kinetoplastid protozoa: Why be normal? Microbes Infect 2003; 5(13):1231-1240.
8. Myler PJ, Audleman L, deVos T et al. Leishmania major Friedlin chromosome 1 has an unusual distribution of protein-coding genes. Proc Natl Acad Sci USA 1999; 96(6):2902-2906.
9. Martinez-Calvillo S, Yan S, Nguyen D et al. Transcription of Leishmania major Friedlin chromosome 1 initiates in both directions within a single region. Mol Cell 2003; 11(5):1291-1299.
10. Martinez-Calvillo S, Nguyen D, Stuart K et al. Transcription initiation and termination on Leishmania major chromosome 3. Eukaryot Cell 2004; 3(2):506-517.
11. Vanhamme L, Pays E. Control of gene expression in trypanosomes. Microbiol Rev 1995; 59(2):223-240.
12. Lodes MJ, Merlin G, deVos T et al. Increased expression of LD1 genes transcribed by RNA polymerase I in Leishmania donovani as a result of duplication into the rRNA gene locus. Mol Cell Biol 1995; 15(12):6845-6853.
13. Perry K, Agabian N. mRNA processing in the Trypanosomatidae. Experientia 1991; 47:118-128.
14. Das A, Zhang Q, Palenchar JB et al. Trypanosomal TBP functions with the multisubunit transcription factor tSNAP to direct spliced-leader RNA gene expression. Mol Cell Biol 2005; 25(16):7314-7322.
15. Schimanski B, Nguyen TN, Günzl A. Characterization of a multisubunit transcription factor complex essential for spliced-leader RNA gene transcription in Trypanosoma brucei. Mol Cell Biol 2005; 25(16):7303-7313.
16. Palenchar JB, Liu W, Palenchar PM et al. A divergent transcription factor TFIIB in trypanosomes is required for RNA polymerase II-dependent SL RNA transcription and cell viability. Eukaryot Cell 2006; 5(2):293-300.
17. Schimanski B, Brandenburg J, Nguyen TN et al. A TFIIB-like protein is indispensable for spliced leader RNA gene transcription in Trypanosoma brucei. Nucl Acids Res 2006; 34(6):1676-1684.
18. Anantharaman V, Aravind L, Koonin EV. Emergence of diverse biochemical activities in evolutionarily conserved structural scaffolds of proteins. Curr Opin Chem Biol 2003; 7(1):12-20.
19. Clayton CE. Life without transcriptional control? From fly to man and back again. EMBO J 2002; 21(8):1881-1888.
20. Klingbeil MM, Motyka SA, Englund PT. Multiple mitochondrial DNA polymerases in Trypanosoma brucei. Mol Cell 2002; 10(1):175-186.
21. Parsons M, Worthey EA, Ward PN et al. Comparative analysis of the kinomes of three pathogenic trypanosomatids: Leishmania major, Trypanosoma brucei and Trypanosoma cruzi. BMC Genomics 2005; 6(1):127.
22. Pink R, Hudson A, Mouries MA et al. Opportunities and challenges in antiparasitic drug discovery. Nat Rev Drug Discov 2005; 4(9):727-740.
23. Fairlamb AH. Chemotherapy of human African trypanosomiasis: Current and future prospects. Trends Parasitol 2003; 19(11):488-494.
24. Lee SH, Stephens JL, Paul KS et al. Fatty Acid synthesis by elongases in trypanosomes. Cell 2006; 126(4):691-699.
25. Albert MA, Haanstra JR, Hannaert V et al. Experimental and in silico analyses of glycolytic flux control in bloodstream form Trypanosoma brucei. J Biol Chem 2005; 280(31):28306-28315.
26. Lakhdar-Ghazal F, Blonski C, Willson M et al. Glycolysis and proteases as targets for the design of new anti-trypanosome drugs. Curr Top Med Chem 2002; 2(5):439-456.
27. Goto S, Nishioka T, Kanehisa M. LIGAND: Chemical database for enzyme reactions. Bioinformatics 1998; 14(7):591-599.
28. Kanehisa M. A database for post-genome analysis. Trends Genet 1997; 13(9):375-376.
29. Kanehisa M, Goto S, Hattori M et al. From genomics to chemical genomics: New developments in KEGG. Nucleic Acids Res 2006; 34(Database issue):D354-D357.
30. Kanehisa M, Goto S. KEGG: Kyoto encyclopedia of genes and genomes. Nucleic Acids Res 2000; 28(1):27-30.
31. Kanehisa M, Goto S, Kawashima S et al. The KEGG resource for deciphering the genome. Nucleic Acids Res 2004; 32(Database issue):D277-D280.
32. Karp PD, Paley S, Romero P. The pathway tools software. Bioinformatics 2002; 18(Suppl 1):S225-S232.
33. Yeh I, Hanekamp T, Tsoka S et al. Computational analysis of Plasmodium falciparum metabolism: Organizing genomic information to facilitate drug discovery. Genome Res 2004; 14(5):917-924.
34. Karp PD, Ouzounis CA, Moore-Kochlacs C et al. Expansion of the BioCyc collection of pathway/ genome databases to 160 genomes. Nucleic Acids Res 2005; 33(19):6083-6089.
35. Ma H, Zeng AP. Reconstruction of metabolic networks from genome data and analysis of their global structure for various organisms. Bioinformatics 2003; 19(2):270-277.

36. Covert MW, Schilling CH, Famili I et al. Metabolic modeling of microbial strains in silico. Trends Biochem Sci 2001; 26(3):179-186.
37. Gaasterland T, Selkov E. Reconstruction of metabolic networks using incomplete information. Proc Int Conf Intell Syst Mol Biol 1995; 3:127-135.
38. Overbeek R, Larsen N, Pusch GD et al. WIT: Integrated system for high-throughput genome sequence analysis and metabolic reconstruction. Nucleic Acids Res 2000; 28(1):123-125.
39. Krieger CJ, Zhang P, Mueller LA et al. MetaCyc: A multiorganism database of metabolic pathways and enzymes. Nucleic Acids Res 2004; 32(Database issue):D438-D442.
40. Rahman SA, Schomburg D. Observing local and global properties of metabolic pathways: 'Load points' and 'choke points' in the metabolic networks. Bioinformatics 2006; 22(14):1767-1774.
41. Rahman SA, Advani P, Schunk R et al. Metabolic pathway analysis web service (Pathway Hunter Tool at CUBIC). Bioinformatics 2005; 21(7):1189-1193.
42. Klamt S, Gilles ED. Minimal cut sets in biochemical reaction networks. Bioinformatics 2004; 20(2):226-234.
43. Lipinski CA, Lombardo F, Dominy BW et al. Experimental and computational approaches to estimate solubility and permeability in drug discovery and development settings. Adv Drug Deliv Rev 2001; 46(1-3):3-26.
44. Hopkins AL, Groom CR. The druggable genome. Nat Rev Drug Discov 2002; 1(9):727-730.
45. Hardy LW, Peet NP. The multiple orthogonal tools approach to define molecular causation in the validation of druggable targets. Drug Discov Today 2004; 9(3):117-126.
46. Zheng CJ, Han LY, Yap CW et al. Therapeutic targets: Progress of their exploration and investigation of their characteristics. Pharmacol Rev 2006; 58(2):259-279.
47. Chen X, Ji ZL, Chen YZ. TTD: Therapeutic Target Database. Nucleic Acids Res 2002; 30(1):412-415.
48. Han L, Cui J, Lin H et al. Recent progresses in the application of machine learning approach for predicting protein functional class independent of sequence similarity. Proteomics 2006; 6(14):4023-4037.
49. Cai CZ, Han LY, Ji ZL et al. SVM-Prot: Web-based support vector machine software for functional classification of a protein from its primary sequence. Nucleic Acids Res 2003; 31(13):3692-3697.
50. Goodsell DS, Morris GM, Olson AJ. Automated docking of flexible ligands: Applications of AutoDock. J Mol Recognit 1996; 9(1):1-5.
51. Jenwitheesuk E, Samudrala R. Identifying inhibitors of the SARS coronavirus proteinase. Bioorg Med Chem Lett 2003; 13(22):3989-3992.
52. Jenwitheesuk E, Samudrala R. Virtual screening of HIV-1 protease inhibitors against human cytomegalovirus protease using docking and molecular dynamics. AIDS 2005; 19(5):529-531.
53. Jenwitheesuk E, Wang K, Mittler JE et al. PIRSpred: A web server for reliable HIV-1 protein-inhibitor resistance/susceptibility prediction. Trends Microbiol 2005; 13(4):150-151.
54. Jenwitheesuk E, Samudrala R. Identification of potential multitarget antimalarial drugs. JAMA 2005; 294(12):1490-1491.

CHAPTER 12

Purine and Pyrimidine Metabolism in *Leishmania*

Nicola S. Carter, Phillip Yates, Cassandra S. Arendt, Jan M. Boitz and Buddy Ullman*

Abstract

Purines and pyrimidines are indispensable to all life, performing many vital functions for cells: ATP serves as the universal currency of cellular energy, cAMP and cGMP are key second messenger molecules, purine and pyrimidine nucleotides are precursors for activated forms of both carbohydrates and lipids, nucleotide derivatives of vitamins are essential cofactors in metabolic processes, and nucleoside triphosphates are the immediate precursors for DNA and RNA synthesis. Unlike their mammalian and insect hosts, *Leishmania* lack the metabolic machinery to make purine nucleotides de novo and must rely on their host for preformed purines. The obligatory nature of purine salvage offers, therefore, a plethora of potential targets for drug targeting, and the pathway has consequently been the focus of considerable scientific investigation. In contrast, *Leishmania* are prototrophic for pyrimidines and also express a small complement of pyrimidine salvage enzymes. Because the pyrimidine nucleotide biosynthetic pathways of *Leishmania* and humans are similar, pyrimidine metabolism in *Leishmania* has generally been considered less amenable to therapeutic manipulation than the purine salvage pathway. However, evidence garnered from a variety of parasitic protozoa suggests that the selective inhibition of pyrimidine biosynthetic enzymes offers a rational therapeutic paradigm. In this chapter, we present an overview of the purine and pyrimidine pathways in *Leishmania*, make comparisons to the equivalent pathways in their mammalian host, and explore how these pathways might be amenable to selective therapeutic targeting.

Nomenclature

Purines and pyrimidines exist as both monomers and polymers. As monomers, they can exist in multiple forms: (1) as free, planar, heterocyclic bases (also called nucleobases); (2) as nucleosides in which an N-glycosidic linkage exists between the one carbon of the ribose or 2'-deoxyribose sugar and either the N9 nitrogen of the purine or N1 nitrogen of the pyrimidine ring; and (3) as nucleotides in which one to three phosphate groups are covalently attached to the five carbon of the ribose ring of a nucleoside. RNA and DNA are nucleotide polymers and informational macromolecules.

*Corresponding Author: Buddy Ullman—Department of Biochemistry and Molecular Biology, Oregon Health & Science University, 3181 SW Sam Jackson Park Rd Portland, Oregon 97239-3098, USA. Email: ullmanb@ohsu.edu

Drug Targets in Kinetoplastid Parasites, edited by Hemanta K. Majumder.
©2008 Landes Bioscience and Springer Science+Business Media.

Purine Metabolism

The discovery that certain pyrazolopyrimidine nucleobases and nucleosides, analogs of naturally occurring purines, are toxic to *Leishmania*, coupled with the obligatory nature of the leishmanial purine salvage pathway, has spawned considerable interest in the purine salvage pathway as a drug target.[1-4] The purine salvage pathway was initially characterized by direct enzyme measurements in parasite lysates and by metabolic incorporation experiments into intact parasites. These studies indicated that the purine salvage pathway of *Leishmania* is complex, entwined, and functionally redundant. More recent molecular and genetic investigations, coupled with the available annotated leishmanial genome sequencing projects, have revealed new details about the purine pathway of *Leishmania* and have emphasized new targets for therapeutic manipulation of the parasite.

Purine Transport in *Leishmania*

Most of our information about the transport of purines and pyrimidines into *Leishmania* has been obtained from studies with *L. donovani* promastigotes, since this form can be easily cultivated in defined medium and is amenable to genetic manipulation.[5] It is clear from experiments involving the supplementation of the cultivation medium with various exogenous purine nucleobases and nucleosides that *Leishmania* are capable of transporting and utilizing any naturally occurring purine nucleobase or nucleoside, including xanthine and xanthosine,[6-9] two purines that cannot be recycled into the nucleotide pool in mammalian cells. The first step in purine salvage is the translocation of preformed purine nucleosides or nucleobases across the parasite plasma membrane. Since nucleosides and nucleobases are hydrophilic and cannot passively diffuse across the lipid bilayer of the plasma membrane, uptake requires specialized translocation proteins or transporters. Through a battery of genetic, molecular, and biochemical studies, as well as comparative in silico analyses on the available *Leishmania* genome databases, four distinct purine transport loci have been identified within the *Leishmania* genome. In *L. donovani* these are designated as *LdNT1-4*, for Leishmania donovani Nucleoside or Nucleobase Transporter. These transporters show limited homology to each other at the amino acid sequence level (~30 % identity), as well as to a family of nucleoside transporters in mammalian cells, termed the Equilibrative Nucleoside Transporters (ENTs, Family 2.A.57 in the Transporter Classification Database, http://www.tcdb.org).[8,9] ENT transporters are distinguished by an overall similarity in predicted topology (11 transmembrane domains and a long intracellular loop between transmembrane segments 6 and 7) and possess a number of conserved or signature residues, located primarily within predicted transmembrane (TM) domains[8] (Fig. 1).

The other class of nucleoside transporter expressed in specialized mammalian cells, designated Concentrative Nucleoside Transporter (CNT) because nucleoside uptake is coupled with the cell's electrochemical gradient[9] (Family 2.A.41 in the Transporter Classification Database, http://www.tcdb.org), appears to be absent in *Leishmania* parasites based upon bioinformatic analyses of the available genome databases.

The functional expression of *Leishmania* nucleoside/nucleobase (*NT*) genes within heterologous expression systems,[13-15] as well as in nucleoside transport-deficient *L. donovani*,[7] have enabled their ligand specificities and kinetic profiles to be assigned. The *LdNT1* locus, which comprises two genes *LdNT1.1* and *LdNT1.2*, and the *LdNT2* gene encode for high affinity nucleoside transporters with nonoverlapping ligand specificities.[13,14] LdNT1 is specific for the purine nucleoside adenosine, as well as for the pyrimidine nucleosides uridine, thymidine and cytidine,[13] while LdNT2 exclusively recognizes the 6-oxopurine nucleosides inosine, guanosine and xanthosine.[7,14] In contrast, *LdNT3* and *LdNT4* appear to encode nucleobase transporters. This observation is surmised largely from parallel work on the related kinetoplastid *L. major*. Studies on LmaNT3, the *L. major* homolog of LdNT3, indicates that it transports the purine nucleobases adenine, hypoxanthine, guanine and xanthine with relatively high affinity.[15] LdNT3, which has not been characterized in detail, mediates the translocation of adenine, hypoxanthine, guanine, and xanthine in yeast (Galazka, Gessford,

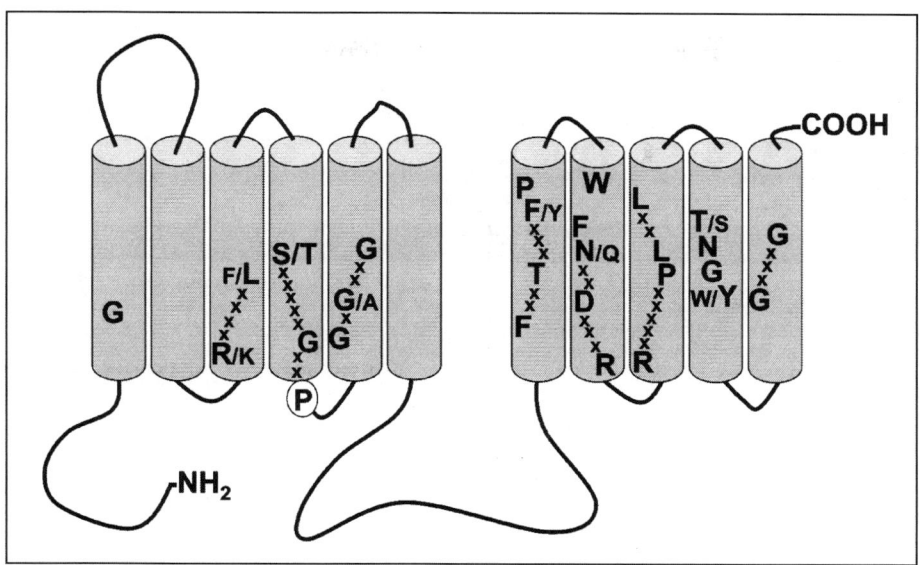

Figure 1. Signature residues delineating the equilibrative nucleoside transporter family. Sequence comparisons were made between the following ENT family members, hENT1,[10] hENT2,[11] hENT3,[12] hENT4,[12] LdNT1,[13] LmaNT1 (LmjF15.1240), LdNT2,[14] LmaNT2 (LmjF36.1940), LdNT3, LmaNT3,[15] LdNT4, LmaNT4 (LmjF11.0550), TbAT1,[16] TbNT2,[17] PfNT1,[18] and TgAT.[19]

and Ullman, unpublished). Likewise, studies on LmaNT4, the *L. major* homolog of LdNT4, suggest that this ENT is an adenine transporter, but exhibits a lower affinity for the nucleobase than LmaNT3 (Ortiz and Landfear, unpublished).

Although the leishmanial NTs share overlapping ligand specificities with mammalian ENTs, there are some notable differences. Electrophysiological measurements in *Xenopus* oocytes with heterologously expressed LdNT1 and LdNT2 demonstrate that unlike their mammalian counterparts, which are equilibrative, these transporters couple a proton for each molecule of ligand translocated.[20] Moreover, unlike most human transporters, which are exquisitely sensitive to the effects of the nucleoside transport inhibitors dipyridamole, dilazep and *S*-nitrobenzyl-4-thioinosine, the leishmanial NTs are largely refractory. The electrogenic nature of the leishmanial permeases and their refractoriness to classical mammalian nucleoside transport inhibitors suggest that the nucleoside translocation mechanism may be discrepant between *Leishmania* and human NTs.

Purine Transporters as Drug Targets in *Leishmania*

Because of the indispensable nature of purine salvage to *Leishmania,* purine transporters may serve as drug targets or as portals through which to introduce toxins. To assess the physiological consequences of loss of nucleoside transport on intact parasites and therefore to establish their suitability as drug targets, a series of null mutants have been created using targeted gene replacement and selection for loss-of-heterozygosity.[7] Thus, we have created Δ*ldnt1*, Δ*ldnt2*, and Δ*ldnt1*/Δ*ldnt2* knockout cell lines. Transport studies on promastigotes and amastigotes of these null mutants demonstrated an absence of adenosine transport in Δ*ldnt1* parasites, a lack of inosine, guanosine, or xanthosine uptake in Δ*ldnt2* cells, and the complete loss of purine nucleoside uptake in Δ*ldnt1*/Δ*ldnt2* mutants.[7] Thus, LdNT1 and LdNT2 appear to be the sole routes for nucleoside uptake in both lifecycle stages of *L. donovani*. Despite these genetic lesions Δ*ldnt1*, Δ*ldnt2*, and Δ*ldnt1*/Δ*ldnt2* cells are all able to grow on any single, natural purine, with the exception that parasites harboring a Δ*ldnt2* lesion cannot utilize xanthosine.[7] More than likely, these discrepancies between transport and growth phenotypes can be ascribed to the extracellular metabolism of purine nucleosides to their respective nucleobases by hydrolytic activities present either in the leishmanial growth medium or on the parasite cell surface. Interestingly, a nonspecific nucleoside hydrolase that recognizes all purine nucleosides appears to localize to the parasite plasma membrane in *L. donovani*[21] and thus may provide a route for purine nucleosides to be converted to nucleobase, ligands for LdNT3 or LdNT4, thereby circumventing the *Δldnt1*, *Δldnt2*, and *Δldnt1/Δldnt2* lesions.

In spite of their lack of nucleoside transport, *Δldnt1*, *Δldnt2*, and *Δldnt1/Δldnt2* cells are as infectious and viable in J774 murine macrophages as wild type parasites during short-term infectivity studies.[7] This is unsurprising given the propensity of these mutants to grow on almost any natural purine. Consequently, our current hypothesis is that pharmacological inhibition of LdNT1 and/or LdNT2 activity is unlikely to cause a loss of viability of *Leishmania* parasites in vivo. However, these transporters may still prove effective for the specific targeting of antileishmanial drugs. Indeed, studies with *Δldnt1* and *Δldnt2* parasites have confirmed that LdNT1 and LdNT2 serve as the primary conduit for uptake of the toxic nucleoside drugs tubercidin, a toxic adenosine analog, and formycin B, a cytotoxic inosine isomer, since *Δldnt1* and *Δldnt2* mutants are highly resistant to these agents.[7]

Structure-Function Studies on Purine Transporters

The successful exploitation of nucleoside transporters in *Leishmania* as a delivery system for toxins is contingent on knowledge of the constraints governing ligand translocation, which are, thus far, largely unknown. However, preliminary structure-function studies of LdNT1 and LdNT2 have yielded some hints about which helices line the ligand translocation channel and have even uncovered particular residues that appear to play a role in ligand discrimination.

Site-directed mutagenesis of charged residues within predicted TM domains in LdNT1 and LdNT2 has revealed roles for some of these residues in transporter function. Mutation of the charged residues Glu94 in TM2 of LdNT1[22] or Asp389 in TM8 of LdNT2[23] results in a properly localized but inactive transporter. These mutant proteins may fail to either bind or translocate ligand; discrimination between these possibilities awaits a cell-free system in which to study these transporters. Intriguingly, a K153R mutation in TM4 of LdNT1 confers inosine transport capability upon the adenosine/pyrimidine nucleoside transporter,[22] and similar changes in ligand specificity have been noted when the corresponding Lys residue in CfNT2, a close relative of LdNT2 from *Crithidia fasciculata*, is mutated (Arendt et al unpublished). Mutation of an amino acid located in a similar position in TM4 of the human transporter hENT1 also affects ligand affinity,[24] suggesting that this portion of TM4 is important in ligand discrimination by ENTs.

Selection for nucleoside transport-deficient mutants by chemical mutagenesis has also highlighted important residues in leishmanial purine transporters. In the TUBA5 cell line, which is null for adenosine/tubercidin uptake,[13] *ldnt1.1* alleles carry inactivating mutations that map to TMs 5 (G183D) and 7 (C337Y) in the mutant ldnt1 proteins.[25] Interestingly, mutation of Gly183 to Ala selectively abrogates uridine transport capability, while adenosine uptake kinetics are unaltered. These data suggest that this particular Gly residue in TM 5 may line the ligand translocation channel and that the uridine and adenosine binding sites are at least partially independent. The FBD5 cell line cannot take up inosine or guanosine[14] and carries a S189L mutation in one of its mutant *ldnt2* alleles, which impairs both activity and efficient cell surface localization of the mutant ldnt2 protein.[26] Mutation of Ser189 to Ala or Thr enables cell surface expression and activity.

Both ldnt1-Gly183 and ldnt2-Ser189 are located within TM5, highlighting this predicted TM domain as critical for ligand transport. Recently, TM5 of LdNT1 has been probed extensively by the substituted cysteine accessibility method.[27] The results of this study indicate that a hydrophilic face of TM5 is accessible to solvent and that most residues on this face are protected from modification in the presence of ligand, indicating that TM5 lines the ligand translocation channel. Notably, both Gly183 and Ser187 (the equivalent of Ser189 from LdNT2) in LdNT1 are located on this hydrophilic face, suggesting that these residues directly affect the flow of ligand through the transporter. Data from chimeras between rat and human ENTs also indicate that TMs 3-6 are important for ligand discrimination,[28,29] suggesting that similar parts of human and protozoan transporters contact ligand.

A structural model has recently been proposed for LdNT2[30] based on threading of ENT primary sequences onto the structure of the glycerol-3-phosphate transporter, a member of the major facilitator superfamily, a family with similar secondary structures to ENTs. In this model, TMs 1, 4, 5, 8, 10 and 11 of LdNT2 line a central channel (Fig. 2). Support for this model comes from the fact that nearly all of the residues found to affect transport activity are located in TMs 4, 5 and 8, all of which line the predicted pore of the ENT. This three-dimensional structural model is therefore consistent with existing mutagenesis results and makes predictions concerning the topological arrangement of helices within the protein that can be experimentally validated, potentially providing valuable structural information for the ENT family.

Another consideration for the therapeutic exploitation of purine transporters is which moieties on the ligand influence its ability to be transported first by mammalian purine transporters and then by the parasite transporters on the surface of the intracellular amastigote. This information will provide constraints for the rational design of effective purine analogs that can gain access to enzymes within the parasite. Little is known about how *Leishmania* and human nucleoside transporters contact ligand; however, the selectivities of human ENT1 vs. LdNT1/ LdNT2 are suggestive. The ubiquitously expressed hENT1 transports all natural purine and pyrimidine nucleosides with moderate affinity[31 32], suggesting that key contacts are likely made to substituents common to all ligands, e.g., the ribose hydroxyls and ring nitrogens within the nucleobase moiety. Indeed, the 3' hydroxyl of ribose and N3 of the pyrimidine heterocycle (similar to N1 of the purine ring) are important for inhibition of uridine uptake by hENT1.[33] In contrast to the broad specificity of hENT1, LdNT1 and LdNT2 discriminate between amino and oxo substituents at the C6 position of the purine ring, implicating this position as particularly vital in forming protein-ligand contacts. Taken together, these observations intimate that drugs that can establish appropriate hydrogen bonds at spatial positions equivalent to N1 and C6 (oxo or amino) in the purine and 3' OH in the ribose ring may be effectively transported into the macrophage and then into *Leishmania* amastigote. In addition, structure/

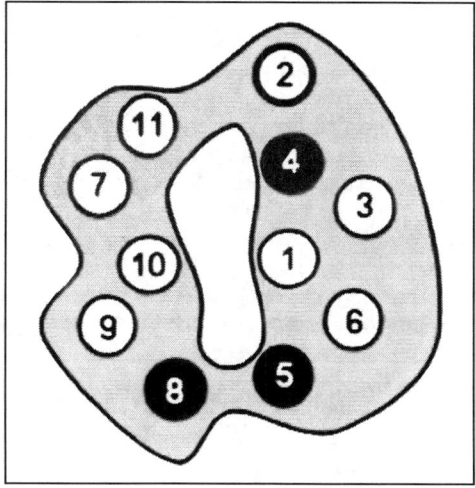

Figure 2. Structural model for arrangement of transmembrane helices in ENT family members. Residues that impact ligand discrimination and transport activity have been identified in helices 2, 4, 5, and 8, three of which are purported to line the water-filled channel (black).

activity studies of TbAT1, an adenosine/adenine transporter from *Trypanosoma brucei*, show that transported drugs need not structurally resemble purine as long as chemically equivalent contacts are provided.[34] Hence, leishmanial nucleoside transporters may provide a gateway for the uptake of diverse cytotoxins.

Purine Salvage Enzymes of *Leishmania*

Leishmania have developed an extensive network of salvage enzymes that enable them to interconvert and metabolize any host purine to the nucleotide level. Through a variety of molecular and biochemical studies, the following purine salvage pathway has been proposed in *Leishmania*.

The pathway is both complex and redundant; host purines can be metabolized to multiple products before proceeding to the nucleotide level. For example, adenosine can either be phosphorylated to produce AMP or hydrolyzed to adenine; adenine can be phosphoribosylated to form AMP or deaminated to hypoxanthine and subsequently phosphoribosylated to IMP.[2] Previous studies on either intact parasites or crude cell lysates implicate adenine phosphoribosyltransferase (APRT), hypoxanthine-guanine phosphoribosyltransferase (HGPRT), and adenosine kinase (AK), as key enzymes within the leishmanial pathway, with the majority of flux proceeding through HGPRT.[2] These studies are complicated, however, by overlapping substrate-specificities of enzymes and the rapid rate of flux through the pathway, which makes the detection of metabolic intermediates difficult. Thus, to examine the role of each of these enzymes a series of mutants were created in *L. donovani*. All of these mutants, including a triple mutant that lacked AK, APRT, and HGPRT activity (*Δhgprt/Δaprt/ak⁻*), were able to proliferate in defined medium containing any natural purine nucleobase or nucleoside.[35,36] This suggests that *L. donovani* in the absence of these activities are able to funnel purine nucleobases through another enzyme in the pathway, xanthine phosphoribosyltransferase (XPRT), and moreover, that none of these activities is essential for cell viability.

Introduction of the *Δxprt* lesion into wild-type cells demonstrated that XPRT is the sole route of xanthine metabolism.[6] However, all attempts to introduce the *Δxprt* mutation into cell lines containing a *Δhgprt* lesion were unsuccessful, even when high levels of exogenous adenine and adenosine were supplied, which are substrates of APRT and AK, respectively.[37] The *Δxprt* mutation could, however, be introduced readily in the presence of *Δaprt* and *ak⁻* mutations (*ak⁻/Δaprt/Δxprt*).[37] These data suggest that all exogenous purines are distilled to substrates for HGPRT or XPRT, and that either HGPRT or XPRT is necessary for the synthesis of nucleotides and thus, for the survival of *L. donovani*.

The inability to create a cell line containing a *Δhgprt/Δxprt* lesion is due to the rapid deamination of adenine to hypoxanthine by the enzyme adenine aminohydrolase (AAH), generating a metabolic dead-end product, which cannot be utilized by the *Δhgprt/Δxprt* cell line. Leishmanial AAH is biochemically distinct from mammalian adenosine deaminase (ADA), although both catalyze a similar reaction that can be pharmacologically inhibited by 2'-deoxycoformycin (dCF).[38,39] Hence we have demonstrated that the conditional lethality of the *Δhgprt/Δxprt* mutation can only be overcome by maintaining these parasites in medium containing both adenine and dCF.[37] Not surprisingly, these mutants are unable to proliferate within murine macrophages in vitro. Currently our laboratory is investigating this *Δhgprt/Δxprt* mutant as a live vaccine candidate.

Purine Salvage Enzymes as Drug Targets in *Leishmania*

The obligatory nature of purine salvage for *Leishmania* highlights its therapeutic potential. However, targeting *Leishmania* purine salvage is complicated since humans and *Leishmania* share several of the same purine metabolizing activities, making the design of *Leishmania*-specific drugs difficult.[4] Any therapeutic strategy that exploits *Leishmania* purine salvage must target either an activity that is unique or essential to the parasite. Through genetic and biochemical analyses we have identified enzymes that are unique to *Leishmania* including XPRT, AAH and possibly various nucleoside hydrolase activities.[6,40] Although none of these enzymes is likely to be essential, they might be exploited to selectively metabolize cytotoxic purine analogs.

Another approach is to inhibit essential, not necessarily unique, enzymes within the *Leishmania* purine salvage pathway. The genetic demonstration that HGPRT and XPRT are essential for purine salvage is a promising lead in the design of novel therapeutics. Even though humans also have HGPRT, differences may exist in the substrate-binding pockets of these enzymes wherein an inhibitor may bind the leishmanial enzyme with higher affinity.[4] This structure-based inhibitor design would be greatly facilitated by high-resolution crystal structures of *Leishmania* HGPRT and XPRT. In addition, since XPRT is also capable of transforming hypoxanthine to the nucleotide level,[6] one might imagine that a single inhibitor could cripple both HGPRT and XPRT. Furthermore, downstream enzymes such as adenylosuccinate synthetase (ADSS) and adenylosuccinate lyase (ASL) (Fig. 3) may also be essential to purine salvage in *Leishmania* and are currently under investigation in our laboratory.

Alternatively, a leishmanial enzyme that also has a human counterpart may be used to selectively metabolize a subversive substrate into a toxic product. This approach is contingent upon

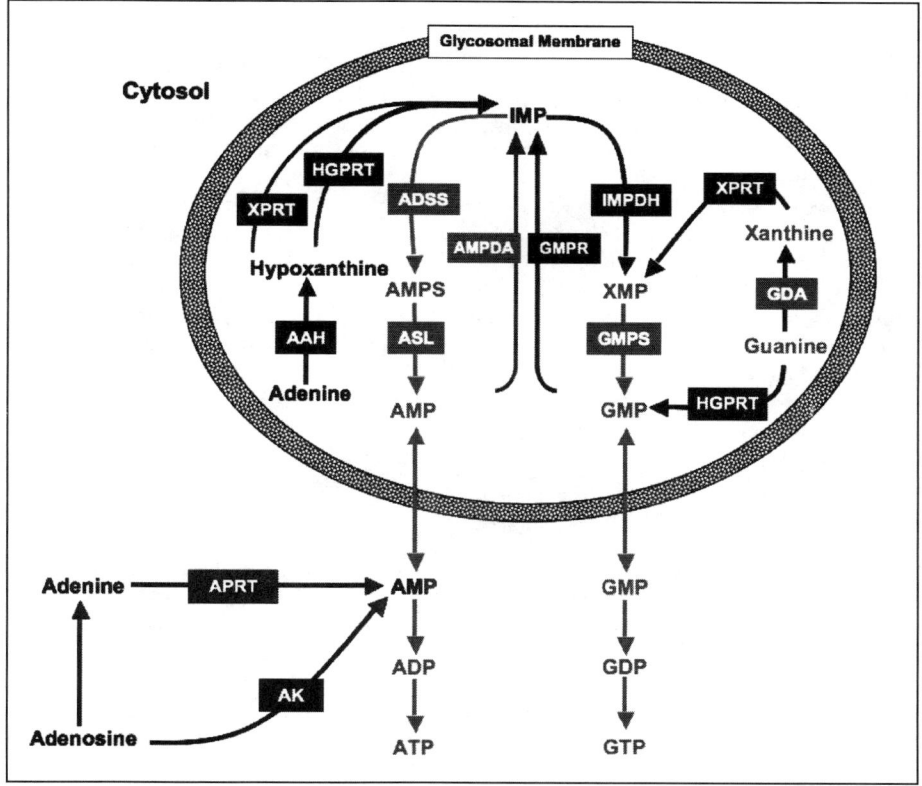

Figure 3. Compartmentalization of the purine salvage pathway of *Leishmania*. Abbreviations are as follows: AAH, adenine aminohydrolase; XPRT, xanthine phosphoribosyltransferase; HGPRT, hypoxanthine-guanine phosphoribosyltransferase; ADSS, adenylosuccinate synthetase; ASL, adenylosuccinate lyase; IMPDH, inosine monophosphate dehydrogenase; GMPS, guanosine monophosphate synthase; GDA, guanine deaminase; AMPDA, adenosine monophosphate deaminase; GMPR, guanosine monophosphate reductase; APRT, adenine phosphoribosyltransferase; AK, adenosine kinase. Enzymes that have been localized are shown in black and those that are predicted to be in the denoted locations are depicted in gray.

structural differences in the substrate-binding pocket. Examples of subversive substrates include the pyrazolopyrimidine analogs of hypoxanthine and inosine such as allopurinol, allopurinol riboside, 4-thiopurinol, 4-thiopurinol riboside, and formycin B, which exert toxicity through incorporation into mRNA.[3,41] *Leishmania* are capable of efficiently metabolizing all five of these pyrazolopyrimidines into nucleotides, whereas humans are not.[41] By far the most studied pyrazolopyrimidine is allopurinol. The clinical efficacy of allopurinol, however, has been disappointing, perhaps as a consequence of competition by natural substrates like hypoxanthine with its metabolism.[4] Nevertheless, the tactic of using purine analogs that are selectively metabolized by the enzymatic machinery of *Leishmania* remains appealing.

The Compartmentalization of Purine Salvage in *Leishmania*

The subcellular milieu of a protein is an important determinant of its specialized function and can also impact drug-targeting paradigms. In *Leishmania* and related parasites, some of the major purine salvage enzymes are compartmentalized within the glycosome, a fuel metabolizing microbody that is unique to these parasites.[42] Glycosomal targeting can be mediated by either a COOH-terminal tripeptide or by a degenerate NH$_2$-terminal signal designated peroxisomal targeting signal-1 (PTS-1) or peroxisomal targeting signal-2 (PTS-2), respectively.[43] It is not known whether proteins lacking these signals can also gain access to the glycosome by "piggy-backing" on proteins with bonafide targeting signals. By direct confirmation of location by immunocytochemical methods (highlighted in black in Fig. 3) or inferred from their primary sequences (highlighted in gray in Fig. 3), the purine salvage pathway is thought to be compartmentalized in *Leishmania* between the glycosome and cytosol as depicted in Figure 3[44,45] (Jardim, A. unpublished; Boitz, J. unpublished).

The fact that key purine salvage components are localized within the glycosome necessitates, perhaps, the translocation of hypoxanthine, guanine, and xanthine across this intracellular membrane. However direct fluorescence studies with episomally expressed LdNT1, LdNT2 and LdNT3, all tagged at the NH$_2$-terminus with green fluorescent protein, suggest that these transporters predominantly localize to the parasite plasma membrane and flagellum[23,25] (Carter et al unpublished). That low levels of these proteins, not detectable in these experiments, are located within the glycosomal membrane, cannot be excluded, however.

The basis for the compartmentalization of the purine salvage pathway within *Leishmania* is not obvious. Glycosomal location may provide a plentiful source of ribose-5-phosphate from the pentose phosphate pathway, which is also sequestered within the glycosome.[42] Ribose-5-phosphate is needed for the synthesis of phosphoribosylpyrophosphate (PRPP), a cosubstrate for all PRT reactions. However, the purine salvage mutant Δ*hgprt*/Δ*xprt* complemented with episomal constructs of *hgprt* or *xprt* that lack a PTS-1 and thus mislocalize to the parasite cytosol, grows robustly in either hypoxanthine or xanthine.[37] This suggests that the glycosomal location of these key components of the purine salvage pathway is not necessary for their function or for parasite viability. Despite the lack of an unambiguous explanation for the unusual organellar distribution of purine salvage enzymes in *Leishmania* and related trypanosomatids, the clear-cut association of therapeutically germane purine salvage enzymes within the glycosome is noteworthy not only from a biological perspective but also from a drug development point of view, as drugs that target HGPRT or XPRT, such as the pyrazolopyrimidine nucleobase analogs[3,46] must traverse the glycosomal membrane to exert their antiparasitic effects.

Pyrimidine Metabolism

Our knowledge of pyrimidine metabolism in *Leishmania* is considerably less detailed than our understanding of purine metabolism, although it has long been known that *Leishmania* is prototrophic for pyrimidines. Previous biochemical studies in combination with more recent genetic and in silico analyses have allowed the de novo biosynthetic and salvage pathways to be defined. Moreover, several lines of evidence (discussed below) suggest that pyrimidine metabolism may present a viable target for pharmacological intervention.

Pyrimidine Biosynthesis in *Leishmania*

In both *Leishmania* and humans, de novo pyrimidine synthesis involves the sequential action of six enzymes: glutamine-dependent carbamoylphosphate synthetase (CPS), aspartate carbamoyltransferase (ACT), dihydroorotase (DHO), dihydroorotate dehydrogenase (DHOD), orotate phosphoribosyltransferase (OPRT) and orotidine monophosphate decarboxylase (OMPDC)[47] (Fig. 4). Despite the obvious similarities in biochemical activities, there are striking discrepancies between the pyrimidine biosynthetic enzymes of *Leishmania* and humans with respect to their organization into multi-functional polypeptides, allosteric regulation, use of cofactors and cellular localization. In humans, a single gene encodes a multifunctional protein encompassing the first three pyrimidine biosynthetic enzymes (designated CAD for CPS-ACT-DHO),[48] while the leishmanial enzymes are encoded by separate genes (http://www.genedb.org; Yates and Ullman, unpublished). Allosteric regulation of CPS governs the rate of pyrimidine synthesis in human cells, and the enzyme is inhibited by both UTP and CTP and stimulated by PRPP.[49] In contrast, the leishmanial CPS is preferentially inhibited by UDP and is unaffected by PRPP.[50] Human DHOD (H-DHOD) is localized to mitochondria and requires ubiquinone as a cofactor,[51] while the *Leishmania* DHOD (L-DHOD) is cytoplasmic[47] and uses fumarate rather than ubiquinone as a cofactor.[52] The last two enzymes of both the mammalian and leishmanial pyrimidine pathways, OPRT and OMPDC, are fused to form a bifunctional protein. However, the arrangements of the two enzymes are reversed; mammalian cells express an OPRT-OMPDC, whereas the leishmanial enzyme is OMPDC-OPRT. Furthermore, OPRT and OMPDC are cytosolic enzymes in mammals, while the presence of a PTS-1 as well as cell fractionation studies[47,53] establish a glycosomal milieu for the parasite OMPDC-OPRT bifunctional protein. This is the only enzyme present within the pyrimidine de novo and salvage pathways that has a glycosomal location.

Another striking feature in *Leishmania spp.* and other trypanosomatids is the organization of the pyrimidine biosynthetic genes[54] (http://www.genedb.org; Yates and Ullman, unpublished). In *Leishmania* all six activities are encoded within a contiguous genomic segment of approximately 25 kilobases. This operon-like grouping of genes involved in a single metabolic pathway is remarkable in view of the fact that genes for virtually every other metabolic pathway are scattered throughout the leishmanial genome.[55]

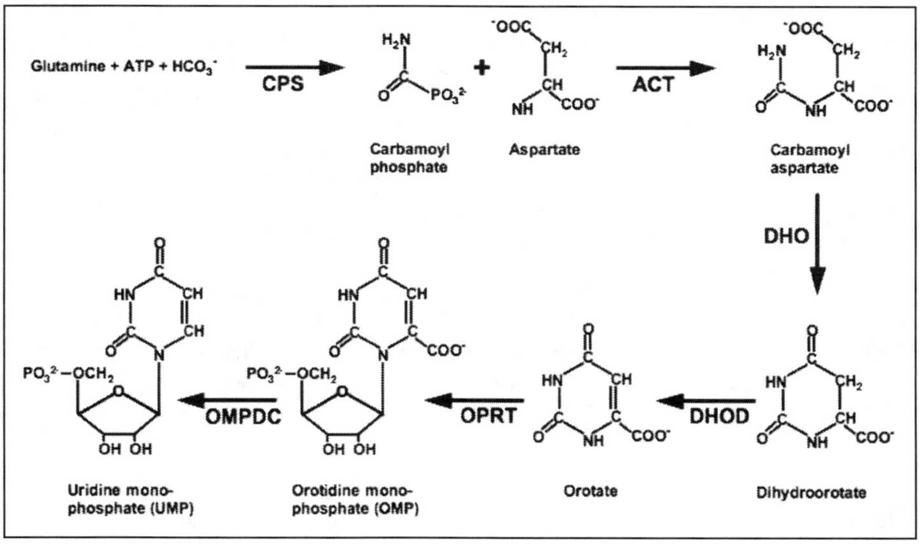

Figure 4. De novo pyrimidine biosynthesis in *Leishmania*. Abbreviations are defined in the text.

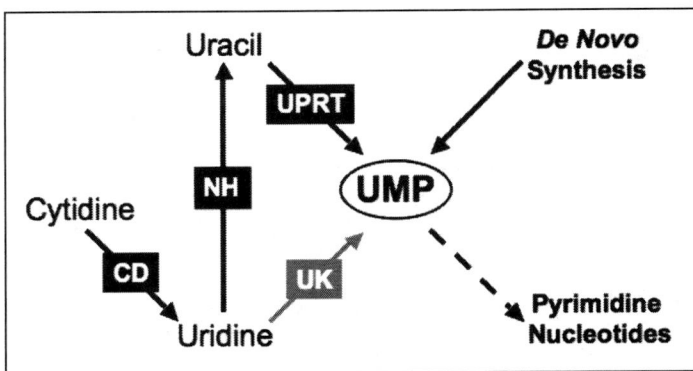

Figure 5. Pyrimidine salvage and nucleotide synthesis in *Leishmania*. Abbreviations are described in the text. Gray text box and arrow indicate that UK enzymatic activity has not been detected. The broken arrow represents the enzymatic steps for conversion of UMP into all other pyrimidine nucleotides as described in the text.

Pyrimidine Salvage and Nucleotide Synthesis in *Leishmania*

The presence of a de novo synthesis pathway implies that pyrimidine salvage may be functionally redundant and potentially less critical to the nutrition of the parasite than purine acquisition. Similar to purine salvage, pyrimidine salvage is initiated by the translocation of pyrimidine nucleosides or nucleobases across the parasite cell membrane via specific transporters. As described earlier, leishmanial NT1 is responsible for the transport of pyrimidine nucleosides.[13] The sole pyrimidine nucleobase transport activity in *Leishmania* has been characterized biochemically in *L. major* (LmU1) and shown to recognize uracil exclusively and with high affinity,[56] but has not yet been cloned.

Upon entry into the parasite, all pyrimidines (except thymidine, cytosine, thymine) are likely converted to uracil, which is phosphoribosylated to UMP by uracil phosphoribosyltransferase (UPRT; Fig. 5).

Leishmania also possess a cytidine deaminase (CD) activity that converts cytidine and deoxycytidine to uridine and deoxyuridine, respectively.[57] Uridine can then be converted to uracil by an inosine-uridine nucleoside hydrolase enzyme[58] and, possibly by other nucleoside hydrolases (NH; Yates and Ullman, unpublished). There is conflicting evidence concerning the existence of a uridine kinase (UK) activity, a more direct route for uridine salvage. UK activity was not detected in *L. mexicana*[47] and unpublished data from this laboratory imply, but do not prove, that UK is also absent in *L. donovani*. However, a gene encoding a putative UK activity is present in the *L. major* genome database. Thymidine is salvaged relatively poorly[59] and is likely directly phosphorylated to TMP by a thymidine kinase (TK), for which putative genes are present in both *L. major* and *L. infantum*.

UMP, the product of both the biosynthetic and salvage pathways, serves as the precursor for all other pyrimidine nucleotides.[60] It is converted to UDP and UTP by nucleotide kinases, UTP is aminated to CTP by CTP synthetase (CTPS), deoxyribonucleotides are synthesized from ribonucleoside diphosphates by ribonucleotide reductase (RR), and thymidylate nucleotides are produced via reductive methylation of dUMP utilizing N^5,N^{10}-methylenetetrahydrofolate as the methyl donor, a reaction catalyzed by a bifunctional dihydrofolate reductase-thymidylate synthase (DHFR-TS) protein.

The Pyrimidine Pathway as a Drug Target in *Leishmania*

A common strategy for drug development is to target enzymes essential for viability. While it is unlikely that pyrimidine salvage is essential, the possibility that de novo pyrimidine biosynthesis is necessary for growth and virulence of *L. donovani* was tested by creating a Δ*cps* knockout via targeted gene replacement (Yates and Ullman, unpublished). Parasites deficient in CPS were auxotrophic for pyrimidines and required supplementation with exogenous pyrimidines for survival and proliferation. Since it is not known if exogenous pyrimidines present in the phagolysosome or mammalian bloodstream are sufficient to circumvent the Δ*cps* lesion, we are currently assessing the ability of Δ*cps L. donovani* to infect macrophages in vitro. In the obligate intracellular parasite *Toxoplasma gondii*, which both synthesizes and salvages pyrimidines, Δ*cps* mutants were avirulent and moreover, elicited a protective immune response.[61] This suggests that the levels of exogenous pyrimidines encountered by the parasite are inadequate to sustain growth. Further evidence supporting the notion that de novo pyrimidine biosynthesis is essential comes from studies of the related trypanosomatid, *Trypanosoma cruzi*, demonstrating that DHOD deficiency is lethal.[62] In *L. donovani*, the glutamine analog acivicin, which inhibits CPS and other enzymes that utilize glutamine,[63] kills both promastigote and intracellular amastigote forms.[64] Similarly, the transition state analogue N-(phosphonoacetyl)-L-aspartic acid (PALA), which specifically targets ATC, inhibits growth of both life cycle stages of *L. donovani*.[65] Taken together, these results intimate that the de novo pyrimidine biosynthesis pathway is essential for virulence and may, therefore, represent a viable therapeutic target.

To date no drugs preferentially target leishmanial components of the pyrimidine biosynthetic pathway. This is likely due to similarities in enzymatic activities shared between *Leishmania* and humans. For example, acivicin and PALA inhibit both human and leishmanial pyrimidine biosynthesis. Similarly, classical inhibitors such as methotrexate, which target leishmanial DHFR-TS, an essential component of thymidylate synthesis,[66] are poor anti-leishmanial drugs because they also inhibit the human enzyme.[67]

In contrast, the unique properties of L-DHOD make it a promising target for drug development. L-DHOD uses fumarate rather than ubiquinone as a cofactor, implying that the substrate-binding pocket differs significantly from H-DHOD. This conjecture is supported by a comparison of the crystal structure of the fumarate-utilizing *Lactococcus lactis* DHOD to that of H-DHOD.[68] That structural differences in DHOD enzymes are amenable to the discovery and design of selective inhibitors is supported by the development of specific inhibitors of other parasite DHOD enzymes, specifically inhibitors of the *P. falciparum* enzyme.[69] Likewise, structural differences in the leishmanial and mammalian bifunctional enzymes that encode OPRT and OMPDC may also be amenable to selective inhibition.

Leishmanial UPRT is a unique enzyme that, while probably not essential, could potentially be subverted to selectively metabolize toxic pyrimidine analogs. Identifying analogs selectively toxic to *Leishmania* could be challenging, as evidenced by the toxicity of the UPRT substrate 5-fluorouracil to both *Leishmania*[56] and humans.[70] While humans lack UPRT, they incorporate 5-fluorouracil into the nucleotide pool via alternative mechanisms; mechanisms that must be bypassed for any uracil analog to demonstrate selective toxicity.

Summary

Purine and pyrimidine metabolism in *Leishmania* offers several metabolic steps for therapeutic intervention. The results of recent molecular and biochemical studies have elucidated the details of the essential pathways for purine and pyrimidine acquisition/biosynthesis, brought to light significant differences between the biochemistry of leishmanial and human enzymes, and allowed the identification of "Achilles heels" in these pathways. These investigations have also uncovered several unique activities, which may be amenable to pharmacological exploitation. These include the leishmanial nucleobase and nucleoside transporters whose novel ligand specificities make them useful as portals through which to target cytotoxins, and the parasite-specific enzymes XPRT, AAH and UPRT whose activities may be plundered to activate prodrugs. Moreover, selective metabolism of prodrugs to the nucleotide level is achievable in *Leishmania* despite overlapping activities with human counterparts. Given the literally thousands of purine and pyrimidine analogs already available for high-throughput screening via cell-based assays or using recombinant enzymes, the future development of purine- or pyrimidine-based antileishmanial agents is quite promising.

References

1. Marr JJ. Purine analogs as chemotherapeutic agents in leishmaniasis and American trypanosomiasis. J Lab Clin Med 1991; 118(2):111-119.
2. Marr JJ, Berens RL, Nelson DJ. Purine metabolism in Leishmania donovani and Leishmania braziliensis. Biochim Biophys Acta 1978; 544(2):360-371.
3. Nelson DJ, Bugge CJ, Elion GB et al. Metabolism of pyrazolo(3,4-d)pyrimidines in Leishmania braziliensis and Leishmania donovani: Allopurinol, oxipurinol, and 4-aminopyrazolo(3,4-d)pyrimidine. J Biol Chem 1979; 254(10):3959-3964.
4. Nelson DJ, LaFon SW, Tuttle JV et al. Allopurinol ribonucleoside as an antileishmanial agent. Biological effects, metabolism, and enzymatic phosphorylation. J Biol Chem 1979; 254(22):11544-11549.
5. Iovannisci DM, Ullman B. Single cell cloning of Leishmania parasites in purine-defined medium: Isolation of drug-resistant variants. Adv Exp Med Biol 1984; 165(Pt A):239-243.
6. Jardim A, Bergeson SE, Shih S et al. Xanthine phosphoribosyltransferase from Leishmania donovani. Molecular cloning, biochemical characterization, and genetic analysis. J Biol Chem 1999; 274(48):34403-34410.
7. Liu W, Boitz JM, Galazka J et al. Functional characterization of nucleoside transporter gene replacements in Leishmania donovani. Mol Biochem Parasitol 2006; 150:300-7.
8. Carter NS, Landfear SM, Ullman B. Nucleoside transporters of parasitic protozoa. Trends Parasitol 2001; 17(3):142-145.
9. Cass CE, Young JD, Baldwin SA et al. Nucleoside transporters of mammalian cells. Pharm Biotechnol 1999; 12:313-352.
10. Griffiths M, Beaumont N, Yao SY et al. Cloning of a human nucleoside transporter implicated in the cellular uptake of adenosine and chemotherapeutic drugs. Nat Med 1997; 3(1):89-93.
11. Griffiths M, Yao SY, Abidi F et al. Molecular cloning and characterization of a nitrobenzylthioinosine-insensitive (ei) equilibrative nucleoside transporter from human placenta. Biochem J 1997; 328(Pt 3):739-743.
12. Baldwin SA, Beal PR, Yao SY et al. The equilibrative nucleoside transporter family, SLC29. Pflugers Arch 2004; 447(5):735-743.
13. Vasudevan G, Carter NS, Drew ME et al. Cloning of Leishmania nucleoside transporter genes by rescue of a transport-deficient mutant. Proc Natl Acad Sci USA 1998; 95(17):9873-9878.
14. Carter NS, Drew ME, Sanchez M et al. Cloning of a novel inosine-guanosine transporter gene from Leishmania donovani by functional rescue of a transport-deficient mutant. J Biol Chem 2000; 275(27):20935-20941.
15. Sanchez MA, Tryon R, Pierce S et al. Functional expression and characterization of a purine nucleobase transporter gene from Leishmania major. Mol Membr Biol 2004; 21(1):11-18.
16. Maser P, Sutterlin C, Kralli A et al. A nucleoside transporter from Trypanosoma brucei involved in drug resistance. Science 1999; 285(5425):242-244.
17. Sanchez MA, Ullman B, Landfear SM et al. Cloning and functional expression of a gene encoding a P1 type nucleoside transporter from Trypanosoma brucei. J Biol Chem 1999; 274(42):30244-30249.
18. Carter NS, Ben Mamoun C, Liu W et al. Isolation and functional characterization of the PfNT1 nucleoside transporter gene from Plasmodium falciparum. J Biol Chem 2000; 275(14):10683-10691.
19. Chiang CW, Carter N, Sullivan Jr WJ et al. The adenosine transporter of Toxoplasma gondii. Identification by insertional mutagenesis, cloning, and recombinant expression. J Biol Chem 1999; 274(49):35255-35261.
20. Stein A, Vaseduvan G, Carter NS et al. Equilibrative nucleoside transporter family members from Leishmania donovani are electrogenic proton symporters. J Biol Chem 2003; 278(37):35127-35134.
21. Cui L, Rajasekariah GR, Martin SK. A nonspecific nucleoside hydrolase from Leishmania donovani: Implications for purine salvage by the parasite. Gene 2001; 280(1-2):153-162.
22. Valdes R, Liu W, Ullman B et al. Comprehensive examination of charged transmembrane residues in a nucleoside transporter. J Biol Chem 2006; 281(32):22647-55.
23. Arastu-Kapur S, Ford E, Ullman B et al. Functional analysis of an inosine-guanosine transporter from Leishmania donovani: The role of conserved residues, aspartate 389 and arginine 393. J Biol Chem 2003; 278(35):33327-33333.
24. SenGupta DJ, Unadkat JD. Glycine 154 of the equilibrative nucleoside transporter, hENT1, is important for nucleoside transport and for conferring sensitivity to the inhibitors nitrobenzylthioinosine, dipyridamole, and dilazep. Biochem Pharmacol 2004; 67(3):453-458.
25. Vasudevan G, Ullman B, Landfear SM. Point mutations in a nucleoside transporter gene from Leishmania donovani confer drug resistance and alter substrate selectivity. Proc Natl Acad Sci USA 2001; 98(11):6092-6097.

26. Galazka J, Carter NS, Bekhouche S et al. Point mutations within the LdNT2 nucleoside transporter gene from Leishmania donovani confer drug resistance and transport deficiency. Int J Biochem Cell Biol 2006; 38(7):1221-9.

27. Valdes R, Vasudevan G, Conklin D et al. Transmembrane domain 5 of the LdNT1.1 nucleoside transporter is an amphipathic helix that forms part of the nucleoside translocation pathway. Biochemistry 2004; 43(21):6793-6802.

28. Sundaram M, Yao SY, Ng AM et al. Chimeric constructs between human and rat equilibrative nucleoside transporters (hENT1 and rENT1) reveal hENT1 structural domains interacting with coronary vasoactive drugs. J Biol Chem 1998; 273(34):21519-21525.

29. Yao SYM, Ng AML, Vickers MF et al. Functional and molecular characterization of nucleobase transport by recombinant human and rat equilibrative nucleoside transporters 1 and 2. J Biol Chem 2002; 277:24938-24948.

30. Arastu-Kapur S, Arendt CS, Purnat T et al. Second-site suppression of a nonfunctional mutation within the Leishmania donovani inosine-guanosine transporter. J Biol Chem 2005; 280:2213-2219.

31. Visser F, Zhang J, Raborn RT et al. Residue 33 of human equilibrative nucleoside transporter 2 is a functionally important component of both the dipyridamole and nucleoside binding sites. Mol Pharmacol 2005; 67(4):1291-1298.

32. Ward JL, Sherali A, Mo ZP et al. Kinetic and pharmacological properties of cloned human equilibrative nucleoside transporters, ENT1 and ENT2, stably expressed in nucleoside transporter-deficient PK15 cells. ENT2 exhibits a low affinity for guanosine and cytidine but a high affinity for inosine. J Biol Chem 2000; 275(12):8375-8381.

33. Chang C, Swaan PW, Ngo LY et al. Molecular requirements of the human nucleoside transporters hCNT1, hCNT2, and hENT1. Mol Pharmacol 2004; 65:558-570.

34. Barrett MP, Fairlamb AH. The biochemical basis of arsenical-diamidine cross-resistance in African trypanosomes. Parasitol Today 1999; 15(4):136-140.

35. Hwang HY, Gilberts T, Jardim A et al. Creation of homozygous mutants of Leishmania donovani with single targeting constructs. J Biol Chem 1996; 271(48):30840-30846.

36. Hwang HY, Ullman B. Genetic analysis of purine metabolism in Leishmania donovani. J Biol Chem 1997; 272(31):19488-19496.

37. Boitz JM, Ullman B. A conditional mutant deficient in hypoxanthine-guanine phosphoribosyltransferase and xanthine phosphoribosyltransferase validates the purine salvage pathway of Leishmania donovani. J Biol Chem 2006; 281(23):16084-9.

38. Kidder GW, Dewey VC, Nolan LL. Adenine deaminase of a eukaryotic animal cell, Crithidia fasciculata. Arch Biochem Biophys 1977; 183(1):7-12.

39. Kidder GW, Nolan LL. Adenine aminohydrolase: Occurrence and possible significance in trypanosomid flagellates. Proc Natl Acad Sci USA 1979; 76(8):3670-3672.

40. Boitz JM, Ullman B. Leishmania donovani singly deficient in HGPRT, APRT or XPRT are viable in vitro and within mammalian macrophages. Mol Biochem Parasitol 2006; 148(1):24-30.

41. Looker DL, Marr JJ, Berens RL. Mechanisms of action of pyrazolopyrimidines in Leishmania donovani. J Biol Chem 1986; 261(20):9412-9415.

42. Parsons M, Furuya T, Pal S et al. Biogenesis and function of peroxisomes and glycosomes. Mol Biochem Parasitol 2001; 115(1):19-28.

43. Opperdoes FR, Szikora JP. In silico prediction of the glycosomal enzymes of Leishmania major and trypanosomes. Mol Biochem Parasitol 2006; 147(2):193-206.

44. Shih S, Hwang HY, Carter D et al. Localization and targeting of the Leishmania donovani hypoxanthine-guanine phosphoribosyltransferase to the glycosome. J Biol Chem 1998; 273(3):1534-1541.

45. Zarella-Boitz JM, Rager N, Jardim A et al. Subcellular localization of adenine and xanthine phosphoribosyltransferases in Leishmania donovani. Mol Biochem Parasitol 2004; 134(1):43-51.

46. Ullman B. Pyrazolopyrimidine metabolism in parasitic protozoa. Pharmaceutical Research 1984; 1(5):194-203.

47. Hammond DJ, Gutteridge WE. UMP synthesis in the kinetoplastida. Biochim Biophys Acta 1982; 718(1):1-10.

48. Carrey EA. Key enzymes in the biosynthesis of purines and pyrimidines: Their regulation by allosteric effectors and by phosphorylation. Biochem Soc Trans 1995; 23(4):899-902.

49. Mori M, Tatibana M. Multi-enzyme complex of glutamine-dependent carbamoyl-phosphate synthetase with aspartate carbamoyltransferase and dihydroorotase from rat ascites-hepatoma cells. Purification, molecular properties and limited proteolysis. Eur J Biochem 1978; 86(2):381-388.

50. Nara T, Gao G, Yamasaki H et al. Carbamoyl-phosphate synthetase II in kinetoplastids. Biochim Biophys Acta 1998; 1387(1-2):462-468.

51. Gero AM, O'Sullivan WJ. Human spleen dihydroorotate dehydrogenase: Properties and partial purification. Biochem Med 1985; 34(1):70-82.
52. Feliciano PR, Cordeiro AT, Costa-Filho AJ et al. Cloning, expression, purification, and characterization of Leishmania major dihydroorotate dehydrogenase. Protein Expr Purif 2006.
53. Hammond DJ, Gutteridge WE, Opperdoes FR. A novel location for two enzymes of de novo pyrimidine biosynthesis in trypanosomes and Leishmania. FEBS Lett 1981; 128(1):27-29.
54. Gao G, Nara T, Nakajima-Shimada J et al. Novel organization and sequences of five genes encoding all six enzymes for de novo pyrimidine biosynthesis in Trypanosoma cruzi. J Mol Biol 1999; 285(1):149-161.
55. Ivens AC, Peacock CS, Worthey EA et al. The genome of the kinetoplastid parasite, Leishmania major. Science 2005; 309(5733):436-442.
56. Papageorgiou IG, Yakob L, Al Salabi MI et al. Identification of the first pyrimidine nucleobase transporter in Leishmania: Similarities with the Trypanosoma brucei U1 transporter and antileishmanial activity of uracil analogues. Parasitology 2005; 130(Pt 3):275-283.
57. Hassan HF, Coombs GH. A comparative study of the purine- and pyrimidine-metabolising enzymes of a range of trypanosomatids. Comp Biochem Physiol B 1986; 84(2):219-223.
58. Shi W, Schramm VL, Almo SC. Nucleoside hydrolase from Leishmania major. Cloning, expression, catalytic properties, transition state inhibitors, and the 2.5Å crystal structure. J Biol Chem 1999; 274(30):21114-21120.
59. LaFon SW, Nelson DJ, Berens RL et al. Purine and pyrimidine salvage pathways in Leishmania donovani. Biochem Pharmacol 1982; 31(2):231-238.
60. Carter NS, Rager N, Ullman B. Purine and pyrimidine transport and metabolism. In: Marr JJ, Nilsen Timothy, Komuniecki, Richard W, eds. Molecular and Medical Parasitology. London: Elsevier Science, 2003:197-223.
61. Fox BA, Bzik DJ. De novo pyrimidine biosynthesis is required for virulence of Toxoplasma gondii. Nature 2002; 415(6874):926-929.
62. Annoura T, Nara T, Makiuchi T et al. The origin of dihydroorotate dehydrogenase genes of kinetoplastids, with special reference to their biological significance and adaptation to anaerobic, parasitic conditions. J Mol Evol 2005; 60(1):113-127.
63. Poster DS, Bruno S, Penta J et al. Acivicin: An antitumor antibiotic. Cancer Clin Trials Fall 1981; 4(3):327-330.
64. Mukherjee T, Roy K, Bhaduri A. Acivicin: A highly active potential chemotherapeutic agent against visceral leishmaniasis. Biochem Biophys Res Commun 1990; 170(2):426-432.
65. Mukherjee T, Ray M, Bhaduri A. Aspartate transcarbamylase from Leishmania donovani. A discrete, nonregulatory enzyme as a potential chemotherapeutic site. J Biol Chem 1988; 263(2):708-713.
66. Titus RG, Gueiros-Filho FJ, de Freitas LA et al. Development of a safe live Leishmania vaccine line by gene replacement. Proc Natl Acad Sci USA 1995; 92(22):10267-10271.
67. Sirawaraporn W, Sertsrivanich R, Booth RG et al. Selective inhibition of Leishmania dihydrofolate reductase and Leishmania growth by 5-benzyl-2,4-diaminopyrimidines. Mol Biochem Parasitol 1988; 31(1):79-85.
68. Norager S, Jensen KF, Bjornberg O et al. E. coli dihydroorotate dehydrogenase reveals structural and functional distinctions between different classes of dihydroorotate dehydrogenases. Structure 2002; 10(9):1211-1223.
69. Baldwin J, Michnoff CH, Malmquist NA et al. High-throughput screening for potent and selective inhibitors of Plasmodium falciparum dihydroorotate dehydrogenase. J Biol Chem 2005; 280(23):21847-21853.
70. Grem JL. 5-Fluorouracil: Forty-plus and still ticking. A review of its preclinical and clinical development. Invest New Drugs 2000; 18(4):299-313.

Index

A

ABC transporter 6, 24, 25, 134
Active transporter 23, 29, 144
Adenosine kinase (AdK) 116, 117, 119-129,
        146, 147
African sleeping sickness 33
Allopurinol 24, 28, 30, 76, 88, 120, 148
American trypanosomiasis 62, 87
Amino acid transporter 25
Anion hole 122
Antifungal drug 61, 64, 66, 67, 75-77
Antiparasitic chemotherapy 116, 119
Apoptosis 4-6, 83, 112, 113
Aquaglyceroporin 23, 24
Aromatic thiocyanate 42, 43
Arsenite 1-5
Azole 2, 51, 61, 63-67, 69, 75-77

B

Benznidazole 61-63, 67, 69, 72-75, 77
Benzylthioinosine (BTI) 121
Brucei 9, 12, 22-25, 28, 34, 36, 38, 40, 48,
        55, 65, 81-85, 87-89, 91, 108, 133-136,
        146

C

Camptothecin (CPT) 5, 106, 108, 110, 111,
        113
Carrier 23-25, 28, 36, 98, 99, 134
Catalytic mechanism 110, 116, 122
Chagas disease 33, 61, 62, 67, 69, 75-77, 87,
        116
Channel 23, 24, 144, 145
Chemotherapy 1-3, 5, 24, 33, 48, 51, 61, 69,
        81-84, 92, 104, 105, 116, 119, 120
Choke-point 133, 135, 137
Compounds 2-4, 22-25, 33-45, 51-54, 61-64,
        66-77, 82-85, 92, 93, 111-113, 116-118,
        133, 136, 138
Cyclophilin (CyP) 127
CYP51 65, 66, 77, *see also* Sterol
        14-demethylase

D

Deacetylases 81-84
Dihydroorotate dehydrogenase (DHOD) 149,
        151
Dinitroaniline sulfonamide 41
Docking 42, 53, 120, 138
Drug
    design 42, 55, 56, 77, 84, 85, 87, 96-98,
            103, 117, 119-121, 133, 136, 138,
            145-147, 151
    development 9, 10, 25, 44, 48, 51, 52,
            55-57, 61, 62, 77, 91, 92, 98, 133,
            135, 138, 148, 150, 151
    efflux 2, 24
    resistance 1-3, 5, 6, 22-25, 48, 88, 105,
            135, 138
    target 1-3, 5, 9, 23, 25, 26, 33-35, 48,
            52-57, 81, 89, 92, 97, 98, 117, 120,
            133-135, 137, 138, 141-143, 146, 150
    target identification 53, 135
    validation 53, 55-57, 117, 135, 138
    uptake 2, 22, 23, 26

E

Equilibrative nucleoside transporter (ENT)
        25, 29, 142-145
Equilibrative transporter 23, 25, 29, 142-145
Ergosterol 43, 55, 64-66, 75, 88, 135
ESAG 134
Etoposide 5, 109, 110

F

Facilitative transporter 23, 26-28
Flavones 113
Formycin B 24, 28, 30, 144, 148

G

Gene deletion 30, 56
Genomics 52, 53, 138
Glucose transporter 23, 25-28
Glycolipid 87, 89, 92-94, 97, 98
Glycoprotein 2, 3, 24, 56, 87, 89, 91-93,
        95-99, 134
Glycosylinositolphospholipid (GIPL) 56, 89,
        91, 92, 94, 95, 97